Identification of CROP and WEED SEEDS

Agriculture Handbook No. 219

Albina F. Musil
Seed Technologist (Botanist)
Seed Branch, Grain Division,
Agriculture Marketing Service

U.S. Department of Agriculture

SCIENTIFIC
PUBLISHERS

Reprint in India, 2015

by

Scientific Publishers (India)

5-A, New Pali Road,
P.O. Box 91
Jodhpur - 342 001 (India)

4806/24, Ansari Road,
Daryaganj
New Delhi - 110 002 (India)

www.scientificpub.com

ISBN: 978-81-7233-828-2

Printed in India

CONTENTS

IDENTIFICATION OF CROP AND WEED SEEDS

By

ALBINA F. MUSIL, *Seed Technologist (Botanist)*

INTRODUCTION

This study was made primarily for the identification of seeds of agricultural plants and field weeds. It is also designed to serve as a guide for the identification of seeds in other related biological sciences in which a knowledge of the gross morphology of seeds is essential. Representative genera and species have been selected to indicate generic relationships that may aid in the identification of unknown species.

Accurate identification of seeds, both crop seeds and weed-seed contaminants, is necessary for the correct labeling of seed moving in commercial channels. This requires skill and good judgment on the part of the seed analyst making the examination. The plant taxonomist may take into consideration all the vegetative parts of a plant from inflorescence to root in making an identification. The seed analyst, on the other hand, must make important decisions on the more limited evidence provided by the fruit or seed alone. In the process of harvesting and cleaning, the fruit or seed may have lost some of its accessory parts or been otherwise damaged, or the degree of maturity may have altered its general appearance.

Seed identification is a specialized field of botany that has been developed over the past 50 years to meet the problem of labeling agricultural and vegetable seeds. There are no organized treatments for the identification of seeds such as the botanical manuals for the identification of plants. The early floras rarely mention the seed. Some of the more recent floras illustrate the seeds, but these seldom show the details necessary for the identification of an unknown seed when considered alone. Papers that have been published on the identification of seeds, while very helpful, are for the most part fragmentary and limited in scope.

Seeds of some kinds of plants are sufficiently distinctive that they are not easily confused with those of other kinds and their identification poses no problems. There are many groups of plants, however, in which seeds of one species may closely resemble those of another species. In some cases one of these may be a crop plant and the other an undesirable or noxious weed. The seed analyst must be able to analyze and evaluate the seed structures of such seeds in relation to those of other similar species. The extensive interregional and international movement of seeds requires a knowledge of species not only in a particular locality but on a worldwide basis.

SCOPE OF STUDY

The text that follows contains descriptive keys and illustrations of seeds of the agricultural crops grown in the United States, and the weed species that may occur as contaminants with crop seeds. One hundred and forty-four genera and 607 species are described. Seeds of well-known crops, such as wheat, barley, rye, corn, flax, and cotton, are not included. Excluded also are a few crops that have only a very limited use as, for example, *Melinis minutiflora*, molessesgrass.

Only a few of the more well-established botanical and horticultural varieties have been included. The number of varieties of cultivated crops is extensive. There are well over a hundred named varieties of such crops as oats and soybeans, and new varieties are continually being developed and put into production. For this reason it is considered impractical to include such a study here.

The number of species of field weeds is almost endless and it is difficult to draw a dividing line as to their relative importance for the entire country. The list of weeds that are defined by law as noxious in the different States may change from year to year as a result of cultural practices or the introduction of new species. An attempt has been made to include in this paper the more important noxious-weed species for all States, some of the more important though not widely distributed species, and weed species that may be indicative of the origin of a crop.

CLASSIFICATION OF PLANTS

In systematic botany closely related plants are combined into groups called *families*, such as the grass family (Gramineae or Poaceae) or the legume family (Leguminosae or Fabaceae).

A family is subdivided into groups showing similar characteristics. Such groups are called *genera* (singular, *genus*). For example, the grass family is divided into many genera, such as *Lolium*, *Poa*, and *Avena*. The genus is further divided into groups which have many characteristics in common. These groups are called *species* (singular, *species*). Using the above illustration, we have *Lolium perenne*, *L. multiflorum*, and others.

In field and garden crops the terms "kind" and "variety" are frequently encountered. As used in this publication, the term "kind" generally refers to a species (alfalfa, *Medicago sativa*) or a botanical variety (chewings fescue, *Festuca rubra* var. *commutata*). In some instances it may refer to a genus, as oats (*Avena*), wheat (*Triticum*), or sweetclover (*Melilotus*).

A "variety," i.e., a cultivated variety, is a subdivision of a "kind" which possesses one or more characteristics (such as morphological, pathological, or chemical) whereby it may be distinguished from other similar forms. Examples: red clover var. Kenland; Kentucky bluegrass var. Merion; radish var. Cherry Belle; beet var. Detroit Dark Red.

THE SEED KEYS

The term "seed" is applied in actual practice to any unit that makes up the bulk of a kind in commercial channels. It is in this sense that the term will be used in this publication. The unit may be a true seed or it may be a true seed with persisting appendages which, botanically, is classified as a fruit. In grasses the unit may be a single floret, an entire spikelet or even a spike, as in certain range grasses, or it may be a hulled caryopsis.

Attention is directed in the seed keys to the most essential features of the seed and only the necessary technical terms have been employed. It sometimes happens that a term which has a specific meaning in describing plants may require a modified interpretation when applied to seeds. The definition of terms in the sense in which they are used in the seed keys may be found in the appended glossary.

As in plant keys, each seed key consists of a series of contrasting features so chosen that in each step one of them will apply to the seed in question and the other will not. The applicable statement is selected in each successive step until the name of the plant to which the seed belongs is reached. If at this stage the seed does not fit in either category, it is probable that an error was made in some previous step in the key and the steps should be reexamined carefully.

It must be borne in mind that there may be a natural range of variation in the seeds of a species, such as seeds from a single pod, a single grass spikelet, or seeds from a single plant. In addition, some parts of the seed structure may have been damaged in harvesting and cleaning and the seed keys are necessarily detailed in order to cover all such conditions. Consequently, not all of the features listed in the keys may be present on every seed, and the determination must be based on a combination of the features present. With experience it is possible to recognize the significant combinations.

An old Chinese proverb "One picture is worth ten thousand words" is especially applicable to the identification of seeds. The accompanying illustrations showing the main diagnostic features of the seeds supplement the descriptions in the seed keys. The illustrations show, in addition to the undamaged seeds, seeds as they may appear after processing. All seeds are shown in various positions to show the principal diagnostic features and, insofar as possible, to indicate the possible range of variation that may be expected within the species. Seeds will doubtless be found that will not look exactly like the illustrations. In order to show all the possible variants, it would be necessary to show many more pictures. In such cases, the basic features shown in the illustrations when combined with the descriptions in the seed key should indicate the significant combination of features whereby the off-type seed can be identified.

Seed measurements are useful mainly in the extreme ranges; that is, in determining whether a seed belongs to a very small-seeded group or to a large-seeded group. Because of natural variation in size within a species, all measurements stated in the seed keys should be considered as approximate only. Shape, unless otherwise stated, refers to the seed as it appears in outline.

Color may be affected by various factors such as climatic conditions, maturity, and aging of seed. Consequently, color may be too variable in some kinds of seeds to be considered a reliable diagnostic feature.

PLANT FAMILIES

Most agricultural seeds belong to either of two large plant groups, (a) the grass family (*Gramineae*) or (b) the legume group (*Leguminosae*), which includes three families: The mimosa family (*Mimosaceae*), senna family (*Caesalpiniaceae*), and pea family (*Fabaceae*). A third family, the mustards (*Cruciferae*), includes many vegetable crops and several field crops. The other families represented in this study include for the most part noxious or objectionable weeds. They are included here for comparison with other similar-appearing seeds.

GRAMINEAE—GRASS FAMILY

STRUCTURAL FEATURES

Spikelets.—The unit of a grass inflorescence is a cluster of florets called a *spikelet* (fig. 1). Spikelets may consist of one or several flowers with a pair of sterile bracts called *glumes* at the base (fig. 1, *a*). The *florets* (fig. 1, *b*) are arranged along a central axis called a *rachilla*, which may or may not be jointed.

Florets.—The individual florets that compose a spikelet consist of a lemma and palea with included flower (pistil and stamens) or the mature ovary (caryopsis or grain).

These structures are shown in figure 2: *Lemma* (*a*), the bract enclosing the grain on the dorsal side; *palea* (*b*), the inner bract enclosing the grain on the ventral side; *rachilla segment* (*c*), a portion of a jointed rachilla which remains attached to the floret at maturity; *callus* (*d*), a smooth thickened layer at the base of a floret where it separates from the rachilla or the main axis of an inflorescence; *keel* (*e*), a sharp fold or angle along the two nerves of a palea or along the midnerve of a lemma; *sinus* (*f*), the space between the margins of a lemma at the base of a floret on the ventral side.

Caryopsis.—With only a few exceptions, the true seed of grasses is fused with the ovary wall (pericarp) and the hilum (fig. 3, *a*) is not a true hilum. This type of fruit is called a *caryopsis*, commonly re-

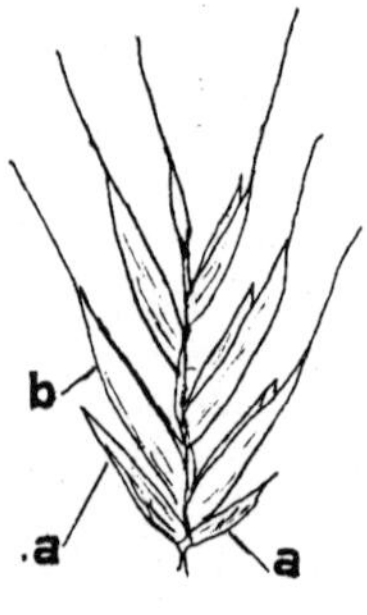

FIGURE 1.—A grass spikelet.

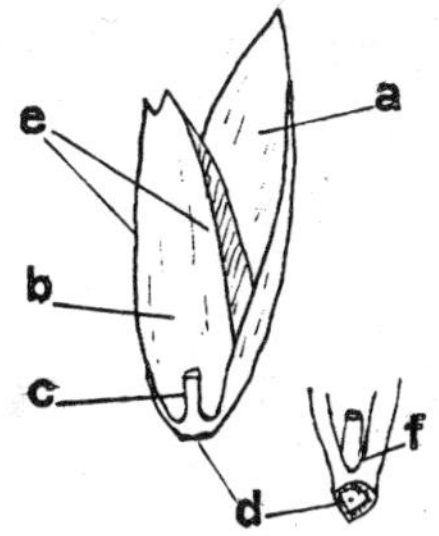

FIGURE 2.—Detail of grass floret.

ferred to as the grain. Characteristic of the grass family, the embryo lies on the outside of an abundant starchy or flinty endosperm (fig. 3, *b*) and is plainly visible near the base of the caryopsis on the dorsal side.

The embryo consists of a scutellum and an embryonic plant. The *scutellum* in most grasses is plainly evident as an oval area on the dorsal side of the caryopsis (fig. 3, *c*). It is a modified cotyledon functioning as a food-absorbing organ. The root-shoot axis (the radicle and plumule of the embryonic plant) usually appears as a slight ridge on the median line of the scutellum (fig. 3, *d*). The embryo side of the caryopsis lies against the lemma and the ventral side lies against the palea of the floret.

SEED UNIT.—The seed unit in grasses may be a caryopsis, a fertile floret, a spikelet, or a spike, depending on the kind or on the method of

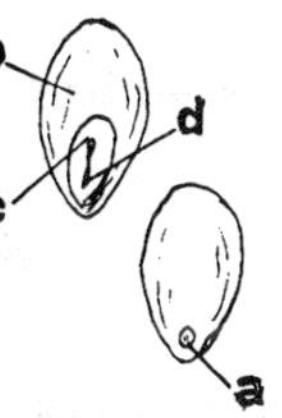

FIGURE 3.—A grass caryopsis.

harvesting and cleaning. The fertilized mature floret, with or without a rachilla segment, is the unit considered the seed in most kinds of grasses in commerce. Some species are completely hulled in processing and the commercial sample consists entirely of naked caryopses. The spikelets of some species do not break apart at maturity and the seed unit may be an entire spikelet. In a few species the entire spike falls from the main axis at maturity and the seed unit in such cases is a spike as, for example, *Bouteloua curtipendula*, side-oats grama.

CLASSIFICATION OF GRASSES

The grass family is divided into two major groups or subfamilies, *Festucoideae* and *Panicoideae*. The subfamilies are further divided into about 14 tribes or subdivisions, not all of which are represented in agricultural seeds.

The rachis or spikelets may be variously modified, but the structures fall into two definite categories. It may not always be possible in the absence of an entire spikelet to place a seed unit in the proper tribe. But with practice it should be possible to narrow the choice to two or three tribes and thus save a time-consuming search through an entire grass manual.

Festucoideae subfamily.—This subfamily is characterized in part by spikelets that are 1- to several-flowered, usually disarticulating above the glumes and between the florets. Eight tribes are represented: (1) *Festuceae*, fescue tribe; (2) *Hordeae*, barley tribe; (3) *Aveneae*, oat tribe; (4) *Agrostideae*, timothy tribe; (5) *Zoysieae*, curly-mesquite tribe; (6) *Chlorideae*, grama tribe; (7) *Phalarideae*, canarygrass tribe; (8) *Oryzeae*, rice tribe.

Panicoideae subfamily.—In this group the 1-flowered spikelets disarticulate below the glumes. The spikelets fall entire, singly, in groups, or together with joints of the rachis. The subfamily is represented by two tribes: (9) *Paniceae*, millet tribe; and (10) *Andropogoneae*, sorghum tribe.

GENERA REPRESENTED IN 10 TRIBES OF GRASSES

Tribe 1. Festuceae, fescue tribe (pp. 9–21)

Bromus (pp. 10–13)
Cynosurus (p. 13)
Dactylis (p. 13)
Distichlis (pp. 13–14)
Eragrostis (p. 14)
Festuca (pp. 14–16)
Glyceria (p. 17)
Poa (pp. 17–20)
Puccinellia (p. 20)
Tridens (p. 21)

Tribe 2. Hordeae, barley tribe (pp. 21–28)

Aegilops (p. 21)
Agropyron (pp. 22–26)
Elymus (pp. 22–26)
Hordeum (pp. 26–27)
Lolium (pp. 27–28)

Tribe 3. Aveneae, oat tribe (pp. 29–34)

Aira (p. 29)
Arrhenatherum (p. 30)
Avena (pp. 30–33)
Danthonia (p. 33)
Deschampsia (p. 33)
Holcus (p. 33)
Koeleria (p. 34)
Sphenopholus (p. 34)
Trisetum (p. 34)

Tribe 4. Agrostideae, timothy tribe (pp. 35–41)

Agrostis (pp. 35–37)
Alopecurus (p. 37)
Apera (p. 38)
Aristida (p. 38)
Calamagrostis (p. 38)
Muhlenbergia (p. 39)
Oryzopsis (p. 39)
Phleum (p. 39)
Polypogon (p. 39)
Sporobolus (pp. 40–41)

Tribe 5. Zoysieae, Curly-Mesquite tribe (p. 41)

Hilaria (p. 41)
Zoysia (p. 41)

Tribe 6. Chlorideae, grama tribe (pp. 42–46)

Beackmannia (p. 42)
Bouteloua (p. 43)
Chloris (p. 43)
Cynodon (p. 44)
Eleusine (p. 44)
Leptochloa (p. 45)
Schedonnardus (p. 46)

Tribe 7. Phalarideae, canarygrass tribe (pp. 46–49)

Anthoxanthum (p. 46)
Ehrharta (p. 47)
Hierochloë (p. 47)
Phalaris (pp. 47–49)

Tribe 8. Oryzeae, rice tribe (p. 49)

Oryza (p. 49)
Leersia (p. 49)

Tribe 9. Paniceae, millet tribe (pp. 49–58)

Axonopus (p. 50)
Cenchrus (p. 51)
Digitaria (p. 51)
Echinochloa (p. 51)
Eriochloa (p. 52)
Leptoloma (p. 52)
Panicum (pp. 53–54)
Paspalum (pp. 54–55)
Pennisetum (p. 55)
Setaria (pp. 56–58)
Stenotaphrum (p. 51)

Tribe 10. Andropogoneae, sorghum tribe (pp. 58-64)

Andropogon (pp. 59–60)
Eremochloa (p. 60)
Hackelochloa (p. 60)
Hyparrhenia (p. 60)
Sorghastrum (p. 61)
Sorghum (pp. 61–64)

SPIKELETS OF REPRESENTATIVE GENERA

There may be certain modifications of the spikelet or rachis within the limits of a tribe, and one example may not represent an entire tribe. Spikelets of some of the more common or representative genera are shown in figures 4–6. The illustrations are designed to show the main structural features and are not drawn to scale. For more complete illustrations of spikelets and glumes, refer to A. S. Hitchcock's Manual of the Grasses of the United States; U.S.D.A. Misc. Publ. No. 200, 1950.

The subfamily *Festucoideae* is represented in the illustrations by tribes 1–8, and subfamily *Panicoideae* by tribes 9–10.

A GUIDE TO THE TRIBES OF GRASSES

1a. Spikelets 2- to several-flowered, usually disarticulating above the glumes and between the florets. (Some exceptions in tribes 2 and 3.)

2a. Seed unit a single floret with attached rachilla segment, the lemma more or less keeled on the back or involute, awnless or awned from the tip or from a bifid apex............................1. *Festuceae*, fescue tribe (p. 9).
3. *Aveneae*, oat tribe (p. 29).
6. *Chlorideae*, grama tribe (in part) (p. 42).

2b. Seed unit a single floret or joint of the inflorescence bearing 1–3 spikelets at each node; lemmas mostly rounded on the back (some exceptions in *Agropyron*); rachilla bristle-like in *Hordeum*.
2. *Hordeae*, barley tribe (p. 21).

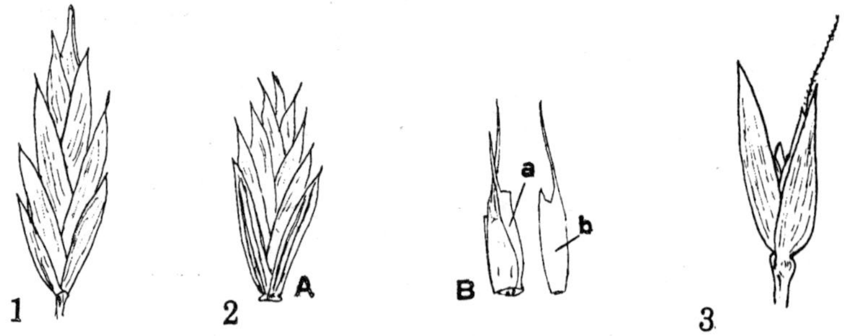

Figure 4.—Representative genera of tribes 1–3. Tribe 1, Festuceae: *Poa*. Tribe 2, Hordeae: *Agropyron* (A) and *Aegilops* (B), showing a spikelet embedded in a joint of the rachis (a) and a detail of the glume (b). Tribe 3, Aveneae: *Avena*.

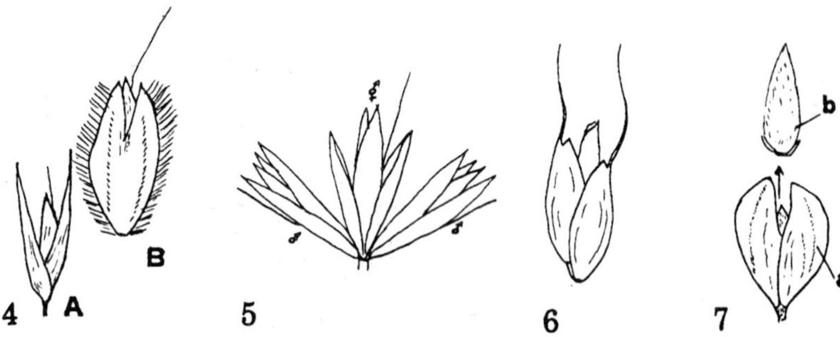

Figure 5.—Representative genera of tribes 4–7. Tribe 4, Agrostideae: *Agrostis* (A) and *Alopecurus* (B). Tribe 5, Zoysieae: *Hilaria*, showing a fascicle of three spikelets, the central spikelet fertile, the two lateral spikelets staminate or sterile. Tribe 6, Chlorideae: *Chloris*, showing the fertile basal floret with the sterile or staminate florets above it. Tribe 7, Phalarideae: *Phalaris*, showing the enlarged glumes (a) and a fertile floret (b) with a pair of persisting scalelike sterile florets at its base.

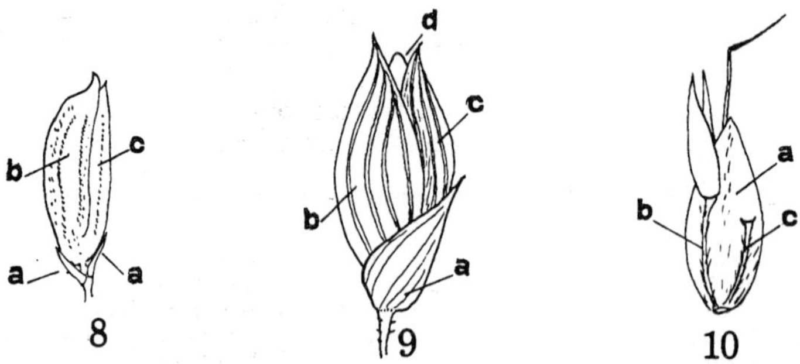

Figure 6.—Representative genera of tribes 8–10. Tribe 8, Oryzeae: *Oryza*, the much reduced glumes (a), the lemma (b), palea (c), Tribe 9, Paniceae: *Panicum*, first glume (a), second glume (b), sterile lemma (c), fertile floret (d). Tribe 10, Andropogoneae: *Sorghum*, glume of fertile spikelet (a), pedicel of sterile spikelet (b), segment of the jointed rachis (c).

1b. Spikelets 1-flowered, with or without attached rudimentary florets; rachilla segments lacking, rudimentary, or prolonged as a bristle. (Exceptions in tribes 4 and 6.)
  3a. Spikelets disarticulate above the glumes.
    4a. Seed unit a single floret with no sterile or rudimentary florets present; lemma rounded on the back------ 4. *Agrostideae*, timothy tribe (p. 35).
    4b. Seed unit a single floret with modified or rudimentary florets attached to the fertile floret; lemma rounded or keeled.
      5a. Rudimentary sterile florets below the fertile floret. 7. *Phalarideae*, canarygrass tribe (p. 46).
      5b. Rudimentary sterile florets above the fertile floret. 6. *Chlorideae*, grama tribe (in part) (p. 42).
  3b. Spikelets disarticulate below the glumes.
    6a. Seed unit a fascicle of 2 to 5 1-flowered spikelets, or single in *Zoysia*, the lemmas and paleas thinner than the glumes. 5. *Zoysieae*, curly-mesquite tribe (in part) (p. 41).
    6b. Seed unit not a fascicle.
      7a. Spikelets strongly compressed laterally.
        8a. Both glumes strongly keeled, dull, sparsely hispid, or both wanting. 8. *Oryzeae*, rice tribe (p. 49).
        8b. First glume lacking, second glume smooth and lustrous, completely enclosing a thin lemma and palea. 5. *Zoysieae*, curly-mesquite tribe (in part) (p. 41).
      7b. Spikelets not compressed laterally.
        9a. Seed unit a spikelet consisting of a thin and tissue-like first and second glume and sterile lemma, and a fertile floret with a hardened lemma and palea------- 9. *Paniceae*, millet tribe (p. 49).
        9b. Seed unit a joint of the inflorescence consisting of a sessile, fertile floret within hardened glumes, a segment of the rachis, and the pedicel (of equal length) of a sterile or rudimentary spikelet. 10. *Andropogoneae*, sorghum tribe (p. 58).

DESCRIPTIONS AND KEYS TO THE GENERA AND SPECIES

1. Festuceae—Fescue Tribe

The spikelets are few to many-flowered, the glumes shorter than the first floret of the spikelet. (See fig. 4, p. 8.) Florets are more or less compressed laterally, keeled on the back or involute, or flattened dorso-ventrally. Lemmas are awnless or awned from the tip or from a bifid apex.

Key to the Genera

1a. Lemmas 3-nerved, the nerves distinct to the base; length 3–4 mm.
  2a. Nerves densely hairy below; florets flattened dorso-ventrally; caryopses obovate-- *Tridens* (p. 21).
  2b. Nerves essentially glabrous; florets mostly flattened laterally; caryopses oblong or oval--------------------------------- *Eragrostis* (p. 14).
1b. Lemmas 3-, 5-, or 7-nerved, at least above, sometimes obscurely so or lacking.
  3a. Lemmas 2½–3 or 4 mm. long.
    4a. Lemmas coarsely granular and hispid across the top, long-pointed with folded apex, awnless or long awned; color sulphur yellow or pale straw. *Cynosurus* (p. 13).
    4b. Lemmas not hispid.
      5a. Nerves faint or lacking; lemmas narrow, rounded on the back, smooth and glabrous or with a few soft hairs across the base- *Puccinellia* (p. 20).
      5b. Nerves distinct, at least at the top, pubescent below; lemmas mostly keeled on the back-------------------------------*Poa* (pp. 17-20).
  3b. Lemmas 5–8 mm. long or up to 16—19 mm. in *Bromus*.
    6a. Lemmas compressed laterally or involute, some exceptions in 7a.
      7a. Lemmas mostly markedly wider than the grain; disc at apex of rachilla tips forward and lies parallel to the palea--------- *Bromus* (pp. 10-13).

7b. Lemmas not wider than the grain; length of floret 5 mm., hairs on keels of palea mostly short, fine, dense.

8a. Lemmas hispid on the midnerve above, nerves fine, faintly evident to the base; width of florets 1–1¼ mm.; caryopses soft, yellowish *Dactylis* (p. 13).

8b. Lemmas glabrous, nerves obscure or lacking; width of florets 2 mm.; caryopses dull, dark brown with minute longitudinal wrinkles. *Distichlis* (p. 13).

6b. Lemmas compressed dorso-ventrally, the margins slightly inrolled.

9a. Nerves 5–7, distinct to the base; lemma short-hispid or coarsely granular; entire palea exposed; hairs on keels minute.... *Glyceria* (p. 17).

9b. Nerves mostly obscure below.

10a. Approximate width of florets 2 mm., entire palea exposed; lemmas glabrous or hispid on the nerves.... *Festuca* (tall fescues) (p. 16).

10b. Approximate width of florets 1 mm., palea partially obscured by the inrolled margins of the lemma... *Festuca* (fine fescues) (p. 14).

BROMUS, BROMEGRASS

The genus *Bromus* includes several valuable forage species, the most important being *B. inermis*, smooth brome. Several annual species occur as weeds in cultivated fields. Some of these are classified as noxious weeds. Some species are objectionable because of the difficulty of cleaning the seeds out of grains and similar crops, or because of some structural features, such as the long sharp awns of *B. rigidus*, which make them dangerous to livestock.

Brome grasses have a characteristic rachilla segment. The back of the rachilla is flat, the front is rounded and bulged out in the upper portion. The callus at the apex tips forward so the plane of the callus tip is parallel to the long axis of the floret. As will be shown later, this type of rachilla segment is also present in a few of the small annual fescues, but other features distinguish them from the bromes.

The seeds of the grasses in this genus are all fairly large and the diagnostic details are visible under magnifications of 6 × or 7 ×. Seeds that have been severely milled may require a magnification of about 20 × to determine the nature of the pubescence. As previously stated, the size of florets may vary depending on their position on the rachilla and the sizes given in the seed key should be considered as approximate only. All measurements are exclusive of the awn.

Fifteen species and certain varieties are described in the seed key. It will be noted that some varieties are distinguishable with accuracy, at least in the bulk. Seeds of other varieties may be too similar to the species to be distinguishable.

Similar-Appearing Species

Plants of the closely related species *B. carinatus* and *B. marginatus* are reported to be quite variable, and some authors consider the two names synonymous. Examination of the seeds of these two species has shown no clearcut features whereby they can be distinguished.

B. japonicus is a widely distributed weed of fields and roadsides, and *B. arvensis* is in limited use as a cultivated crop. Seeds occurring as a contaminant in crop seeds are most likely to be *B. japonicus*. Plants of the two species are separated on rather arbitrary and variable characters, and in some cases they may be indistinguishable. Seeds used for the study of these two species were taken from mature plants grown in different parts of the country. It was observed that seed in the bulk appears to be distinguishable, but not all individual seeds of *B. arvensis* conform to the characteristics set forth in the seed key.

Seeds of *B. japonicus* when occurring with crop seeds are usually tightly rolled and the tips of the fragile lemmas are broken off. The small size of the seed and the delicate texture of the lemma serve to distinguish it from the more robust *B. commutatus* with which it is sometimes confused.

Accurate identification of seeds of *B. secalinus* and *B. commutatus* is important because one or both species are considered noxious weeds in many States. Seeds of the two species when fully matured are readily distinguishable, but seeds that were harvested before the grain had filled out completely may superficially appear similar. Identification of such underdeveloped seeds has to be made without the aid of a fully developed caryopsis, and a critical examination must be made of the lemma-palea relationship.

The seed key is arranged in three sections. Section 1: The lemmas are flattened dorso-ventrally throughout. Section 2 (p. 11): Lemmas are sharply compressed laterally, completely obscuring the palea. Section 3 (p. 12): The lemmas are more or less involute and keeled just above the base.

Key to *Bromus*

Section 1

Lemma flattened dorso-ventrally throughout.

Lemmas 3--5-nerved, short-pubescent near the base, length 9–12 mm., width 2 mm.; awn 1–2 mm. long; palea narrower than the caryopsis, the hairs on keels very short, fine, close-spaced; rachilla bristly-pubescent; caryopsis thin and flat ________ *B. inermis*, smooth brome (pl. XXV, 116).

Section 2

Lemma sharply compressed laterally, completely obscuring the palea; awns 2–3 mm. long, or longer in 1*a*.

1*a*. Lemma in side view slender throughout or broader about midway, the nerves obscure; sparsely appressed-pubescent and minutely hispid, rough to the touch; length 15–16 mm.; width 1½ mm.; rachilla granular or hispid.
B. marginatus, mountain brome (pl. XXV, 118).
B. carinatus, California brome.

1*b*. Lemma in side view broadest below the middle, the nerves evident at least on upper portion; rachilla glabrous or short-pubescent.

2*a*. Lemma short-pointed, essentially glabrous and smooth, occasionally sparsely pubescent with short, rather soft hairs, the nerves coarse, obscure below; length 15–16 mm., width 3 mm.
B. catharticus, rescuegrass (pl. XXV, 117).

2*b*. Lemma long-pointed, minutely hispid and somewhat rough to the touch.

3*a*. Nerves of lemma obscure below or the marginal nerves evident to the base, the midnerve hispid, lemma sparsely hispid especially on upper portion and along the margins; length 18–19 mm., width 3–3½ mm.
B. catharticus var. Chapel Hill.

3*b*. Nerves of lemma mostly evident to the base.

4*a*. Pubescence on lemma mostly confined to the nerves, the nerves fine, each bearing a row of short prickles; margins of lemma prominently winged; length 18–20 mm., width 3–3½ mm.
B. catharticus var. Prairie brome.

4*b*. Pubescence on lemma not confined to the nerves, the nerves coarse; lemma minutely hispid and granular, somewhat rough to the touch; length 14–15 mm., width 1½–2 mm. ___ *B. stamineus* var. Harlan brome.

Section 3

Lemma more or less involute and keeled just above the base.

1a. Length of lemma about 8 times the width or longer, the apex terminating in two long, hyaline teeth; awns 1–2 times the length of the lemma; callus wide (vertically), pointed; caryopsis thin.

2a. Lemma copiously soft-pubescent, the hairs long and lax across the top and along the sides; palea concavity shallow, the hairs on the keels fine, wide-spaced, with shorter hairs between; callus with a few short hairs at each end.

3a. Lemma only slightly wider than the caryopsis in the lower half, the palea well exposed; length 8–12 mm.
B. tectorum, downy chess (pl. XXV, 110).

3b. Lemma markedly wider than the caryopsis in lower half, inrolled and partially obscuring the palea below; length 12–16 mm.
B. rubens, foxtail chess.

2b. Lemma short-hispid, rough to the touch; palea concavity deep, V-shaped, the palea mostly obscured by the inrolled lemma.

4a. Callus yellowish, polished, glabrous or with a short tuft of hairs at each end; hairs on keels of palea long, fine, wide-spaced; length 13–15 mm.
B. sterilis, barren chess (pl. XXV, 111).

4b. Callus dull, with a whitish pubescence across the top; hairs on keels of palea mostly short, coarse, wide-spaced; length 25–30 mm.
B. rigidus, ripgutgrass.

1b. Length of lemma about 4–5 times the width or less, the apex obtuse, sometimes cleft in *mollis*; awns equal to the lemma or shorter, sometimes lacking in *secalinus;* callus not wide (vertically) or pointed.

5a. Lemma, palea, and caryopsis equal in length or nearly so, the lemma about the same width as the caryopsis in the upper ¼ and only slightly exceeding it below; length 7–8 mm., width 1½–2 mm.; caryopsis thick, strongly compressed laterally............. *B. secalinus*, chess or cheat (pl. XXV, 114).

5b. Lemma, palea and caryopsis not equal in length; lemma wider than the caryopsis in the upper ⅓; caryopsis thin.

6a. Caryopsis strongly compressed laterally, the concavity deeply U-shaped throughout; lemma delicate in texture, much wider than the caryopsis and often strongly involute, length 7–8 mm.; caryopsis very thin, 5½–6 mm. long, 1 mm. wide.

7a. Palea markedly shorter than the lemma.
B. japonicus, Japanese chess (pl. XXV, 112).

7b. Palea mostly equal to or slightly shorter than the lemma.
B. arvensis, field brome.

6b. Caryopsis not strongly compressed laterally, the concavity shallow in lower portion and flattened out toward the top, especially in *mollis*, length 6–6½ mm., width 1½–2 mm.; lemma 8–9 mm. long.

8a. Lemma thin and tissue-like often with minute transverse wrinkles; nerves fine lines, usually evident to the base; the palea well exposed.

9a. Lemma minutely short-pubescent, the hairs soft.
B. mollis, soft chess (pl. XXV, 115).

9b. Lemma minutely scabrous, slightly rough to the touch.
B. racemosus, smooth chess.

8b. Lemma stiff, essentially glabrous, the coarse midnerve evident to the base, other nerves obscure below; the broad lemma usually inrolled, obscuring much of the palea.
B. commutatus, hairy chess (pl. XXV, 113).

Hulled Caryopses

Hulled caryopses of *Bromus secalinus*, *B. commutatus*, and *B. japonicus* may occur in crop seeds, occasionally in considerable number. It is important that they be identified since a noxious weed is involved. With practice it should be possible to recognize the species even though the caryopses are slightly damaged. The following tabulation, used in connection with the illustrations (pl. XXV, 112–114), may serve as a guide for this study.

Caryopses of 3 species of *Bromus* compared

Species	Length (*mm.*)	Width (*mm.*)	Shape
B. secalinus	5½–6	1½–2	The sides thick, strongly inrolled but flattened out at the extreme tip, a sharp V-shaped crease down the middle; the sides taper gradually at the base to a short-pointed embryo.
B. commutatus	5½–6	1½–2	The sides thin, only slightly inrolled, the concavity broad and shallow; the sides narrow abruptly at the base to a short-pointed embryo, producing minute lobes just back of the radicle tip.
B. japonicus	5½–6	1	The sides paper thin, only slightly inrolled, the concavity shallow; the sides taper gradually at the base to a pointed embryo.

CYNOSURUS, DOGTAIL

Two species of *Cynosurus* are of agricultural interest. *C. cristatus*, crested dogtail, is sometimes used in grass mixtures for meadows, and *C. echinatus*, bristly dogtail, may occur as a contaminant with field crops.

The two species have several features in common. Lemmas are rounded on the back, coarsely granular and with short hispid hairs across the top and along the sides. Nerves are fine lines, usually obscure below, and the callus is a fine smooth rim. Paleas are flat with inturned keels, the apex obscured by the folded tips of the lemmas.

Key to *Cynosurus*

1a. Floret sulphur yellow with whitish hispid hairs across the top of the lemma; callus lies in a vertical position; rachilla stout, with a wide disk at the apex; lemma 3½–4 mm. long, 1 mm. or less wide; awn 1 mm. long.
C. cristatus, crested dogtail (pl. XXV, 123).

1b. Floret pale straw-color; callus lies in an oblique position; rachilla slender, the apex scarcely expanded; lemma 5–7 mm. long, 1¼–1½ mm. wide; awn 10–15 mm. long ------------------- *C. echinatus*, bristly dogtail (pl. XXVI, 124).

DACTYLIS, ORCHARDGRASS

D. glomerata, orchardgrass (pl. XXVI, 125).—The lemma is compressed laterally, the apex commonly bent off-center, the keel nerve ciliate toward the top. The caryopsis is soft and markedly shorter than the palea. Lemmas are 5–7 mm. long or longer, the awn about 2 mm. long. Commercial seed sometimes carries florets with a sterile terminal floret persisting.

DISTICHLIS, SALTGRASS

The hulled caryopses of *Distichlis* are found with certain crop seeds. Lemmas and paleas are loose-fitting on the caryopses and the intact florets rarely occur in processed seed. Two species of rather wide distribution are described.

Caryopses are keeled on the dorsal side and broad on the ventral side. The color is a dull, light brown, the surface minutely wrinkled. Remnants of the two style branches appear as short beaks or prongs at the top.

600705 O—62——2

D. stricta, desert saltgrass (pl. XXVI, 134).—Caryopses are 3–4 mm. long and terminate in a long, two-pronged beak. Occasional caryopses are longer and narrower than those shown in the illustration, and the beak is correspondingly longer. When found in processed seed the prongs of the beak may be partially broken away, but the size of the caryopsis and the remnant of the beak will usually distinguish it from *D. spicata*.

D. spicata, seashore saltgrass (pl. XXVI, 133).—Caryopses are about 2 mm. long and terminate in two minute prongs at the apex. The apex is not prolonged into a long beak as in *D. stricta*.

ERAGROSTIS, LOVEGRASS

Seven of the more important species of *Eragrostis* are described. Two of these are of agricultural importance at the present time; *E. curvula*, weeping lovegrass, and *E. trichodes*, sand lovegrass. The other species are more limited in use or of experimental interest.

The lemmas are 3-nerved, the lateral nerves sometimes obscure. Lemmas and paleas are rather delicate and easily detached so that processed seed consists almost entirely of hulled caryopses.

Key to the Caryopses of *Eragrostis*

1a. Caryopses oval, light brownish yellow, smooth or faintly striate; embryo area black.
  2a. Embryo area oblong with a distinct rim; length of caryopsis 1¾–2 mm., width 1 mm. or less.----*E. curvula*, weeping lovegrass (pl. XXVII, 147).
  2b. Embryo area obtusely triangular, not distinctly rimmed; length of caryopsis 1 mm. or less, width ½ mm.
    E. lehmanniana, Lehmann lovegrass (pl. XXVII, 148).
1b. Caryopses, including scar area, light brown or dark reddish brown.
  3a. Ventral side with a deep longitudinal cavity down the middle; caryopses broadly oblong, surface reticulate.
    4a. Caryopsis 1–1½ mm. long, 1 mm. or more wide; reticulum faint.
      E. trichodes, sand lovegrass (pl. XXVII, 149).
    4b. Caryopsis ½–¾ mm. long, ½ mm. wide; reticulum distinct.
      E. capillaris, lacegrass (pl. XXVII, 144).
  3b. Ventral side without longitudinal cavity.
    5a. Caryopsis rotund, broadly oval, apex and base pointed, surface faintly reticulate; length and width about equal, ¾—1 mm.
      E. cilianensis, stinkgrass (pl. XXVII, 145).
    5b. Caryopsis not rotund or pointed.
      6a. Surface finely striate; caryopsis oblong, length 1 mm. or more, width ½ mm.---------*E. chloromelas*, Boer lovegrass (pl. XXVII, 146).
      6b. Surface neither striate nor reticulate; caryopsis strongly arched on ventral side, length 1–1½ mm. width ½ mm.------------*E. tef*, teff.

FESTUCA, FESCUE

The genus *Festuca* is large and widespread. Many of the species are valuable range grasses, but comparatively few are of agronomic interest either as cultivated crops or field weeds.

Nine species are described in the seed key. Four of these species are cultivated crops: *F. rubra*, *F. ovina*, *F. arundinacea*, *F. elatior*, including several horticultural varieties. *F. capillata* is not being grown here but occurs as an incidental crop seed in some imported seed lots. The other five species may occur as contaminants in crop seeds.

Similar-Appearing Species

Some intergrading in lemma pubescence may be expected in *F. pacifica* and *F. myuros*, and some seeds may be indistinguishable. *F. myuros* is the more widely distributed of the two species.

It will be observed in the illustrations (pl. XVI, 39, 40) that the primary or basal floret in the spikelets of *F. megalura* lacks the long criss-cross hairs on the margins of the lemma. These basal florets could be confused with *F. myuros*.

As shown in the illustrations, the rachillas of the species under *2a* and *2b* in the seed key are similar to those of *Bromus*. (Refer to the description on p.10, par.2.)

Two varieties of tall fescue are grown extensively: *F. arundinacea* var. Alta and *F. arundinacea* var. Ky.-31. Plants of the two varieties are reported to be very similar and the seeds are doubtfully distinguishable. In general, the seeds of the northern-grown var. Alta are lighter in color than those of the southern-grown var. Ky.-31.

There are several horticultural varieties of *Festuca rubra* in cultivation, the more important of which are Illahee, Pennlawn, Rainier, and one botanical variety, *F. rubra* var. *commutata*, Chewings fescue. The variety Rainier is usually distinguishable by the larger size of the seed. Seeds of Illahee, Pennlawn, and Chewings are too similar to be identified with certainty in all cases.

Seeds of *Festuca ovina*, sheep fescue, and *F. duriuscula*, hard fescue, are very similar. Seeds of hard fescue tend to be more slender and longer pointed than sheep fescue, and the palea appears to be more opaque. With experience, it may be possible to identify bulk samples but the individual seeds appear to be indistinguishable.

The comparative size and shape of 10 kinds of fine-leaved fescues are shown in plate XV, 38.

Key to *Festuca*

1a. Lemma linear, awns about twice the length of the lemma; rachillas thread-like, the apex scarcely expanded. (See Bromus, p. 10, par. 2.)

2a. Edges of lemma with long stiff cilia toward the top, except on the primary florets; lemma scabrous and rough to the touch, the inrolled margins leave only a narrow strip of palea exposed; length 5–6 mm., width ½ mm. or more.
F. megalura, foxtail fescue (pl. XVI, 40).

2b. Edges of lemma devoid of cilia.

3a. Paleas folded at the tip.

4a. Margins of lemma inrolled, only a narrow strip of palea exposed; sparsely short-pubescent at the top and along the sides, the awn slender; length 5 mm., width ½ mm.
F. myuros, rattail fescue (pl. XVI, 39).

4b. Margins of lemma extend to keels of palea, the palea well exposed below; lemma stiff, scabrous across the top and along the sides, terminating in a coarse hispid awn; length 5½–6 mm., width ¾ mm.
F. pacifica, Pacific fescue.

3b. Palea flattened out at the tip, the palea well exposed; lemma sparsely short-pubescent across the top and along the sides; length 6–7 mm., width ¾–1 mm.......... *F. dertonensis*, big rattail fescue (pl. XVI, 42).
(*F. bromoides* of Auth. not L.)

1b. Lemma lanceolate or elliptic; awns vary from about half the length of the lemma to awnless; rachilla cylindric, the apex expanded into a disk.

5a. Lemmas broadly elliptic, length 3 mm., width 1 mm., glabrous, awnless or with an awn-point; rachilla disk not prominent.
F. capillata, hair fescue (pl. XVI, 41).

5b. Lemmas not broadly elliptic.

6a. Seeds narrow with a deep V-shaped cavity, only a narrow strip of palea visible.

7a. Seeds relatively small, length 3½–4 mm., width ¾ mm. or less; lemma sparingly hispid across the top and along the sides; awns about half the length of the lemma; rachilla short, stout, with a prominent apical disk.................. *F. octoflora*, sixweeks fescue (pl. XV, 38D).

7b. Seeds larger than above, length 4–7 mm. or longer, width 1–1½ mm., pubescence and awns various.

8a. Keels of palea converge abruptly above the grain as if pinched together; lemma and palea sparsely pubescent across the top and glabrous below............... *F. ovina*, sheep fescue (pl. XVII, 45).
F. duriuscula, hard fescue.

8b. Keels of palea converge gradually toward the top, the deeply notched apex folded. *F. rubra*, red fescue and vars. (pl. XVII, 43, 44, 46).

6b. Seeds broadest about midway, tapering about equally to a pointed apex and base; palea concavity broad and shallow.

9a. Lemma essentially glabrous, semi-glazed, awnless, the apex and base short-pointed; palea glabrous and lustrous.
F. elatior, meadow fescue (pl. XVIII, 47).

9b. Lemma coarsely granular, hispid on the nerves, long-pointed; palea granular, dull............ *F. arundinacea*, tall fescue (pl. XVIII, 48).

Ryegrass, Tall Fescue and Meadow Fescue Compared

Seed samples of tall fescue (*Festuca arundinacea*) and meadow fescue (*F. elatior*) sometimes contain an admixture of ryegrass (*Lolium perenne* or *L. multiflorum*). Small seeds or terminal florets of ryegrass are easily overlooked in such mixtures. Terminal florets of all three kinds usually have long slender rachilla segments which do not have a characteristic shape, and the identification must be made entirely on other features. Mixtures of tall fescue and meadow fescue are sometimes difficult to separate if the seed has been severely milled.

Figure 7 shows the characteristic shapes of the florets and rachilla segments of the three kinds. The tabulation that follows describes other structural features whereby the three species can be distinguished.

Structural Features of Ryegrass, Tall Fescue, Meadow Fescue

Seed structure	*Lolium* spp. (pl. XIX, 51)	*F. arundinacea* (pl. XVIII, 48)	*F. elatior* (pl. XVIII, 47)
Shape	Oblong	Elliptic, long-pointed.	Elliptic, short-pointed.
Rachilla segment.	Flat, wide, strap-like, the apex not expanded.	Round, slender, the apex expanded into a disk.	As in *arundinacea*.
Callus	A fine clean-cut rim, narrow (vertically).	Thick, often with an indentation in the upper edge, wide (vertically).	As in *arundinacea*.
Lemma	Granular, sparingly pubescent or glabrous.	Coarsely granular, scabrous on nerves and along the margins.	Smooth, as if glazed, glabrous.
Awns	Mostly lacking in *perenne*, long or short in *multiflorum*.	Present	None.
Palea	Smooth and lustrous toward the top.	Coarsely granular, dull.	Smooth and lustrous toward the top.

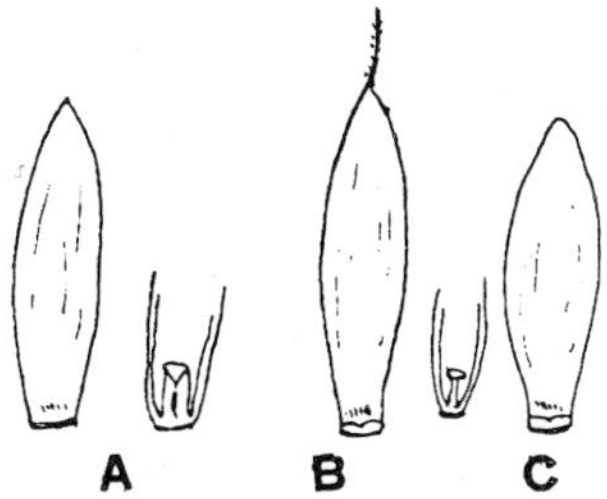

FIGURE 7.—Shapes of florets and rachilla segments of (A), ryegrass (*Lolium perenne* and *L. multiflorum*); (B), tall fescue (*Festuca arundinacea*); and (C), meadow fescue (*F. elatior*).

GLYCERIA, MANNAGRASS

Twenty species of *Glyceria* are reported for the United States. Three of the more widely distributed species, which may occur with crop seeds, are described. Florets of the genus are characterized by the coarsely granular and 7-nerved lemmas. The paleas are equal to or slightly exceed the lemmas and are completely exposed. Hairs on the keels of the paleas are close-spaced and so small they are scarcely visible at a magnification of 20 ×.

Key to *Glyceria*

1a. Florets lance-shaped, long-pointed, 5–6 mm. long, 1½ mm. wide; lemmas granular, short-hispid and rough to the touch, hyaline along the margins and across the top; caryopses flattened dorso-ventrally, dark brown, dull or lustrous and roughened by coarse anastamosing lines.
G. fluitans, mannagrass (pl. XXVII, 152).

1b. Florets not long-pointed, lemmas granular and minutely hispid on the nerves, the margins not hyaline.

2a. Floret oval, the apex and base obtuse, length 2½ mm., width ¾ mm.; caryopsis flattened dorso-ventrally, dark brown, smooth and glossy.
G. grandis, American mannagrass (pl. XXVII, 153).

2b. Floret obovate, thickened above the middle, the lemma and palea spread apart by the thick caryopsis, length 1½ mm., width ¾ – 1 mm.; caryopsis top-shaped, dark brown, smooth and glossy.
G. striata, (*Panicularia nervata*) fowl mannagrass (pl. XXVII, 154).

POA, BLUEGRASS

Sixty-nine species of *Poa* are recorded for the United States. Thirteen species and six varieties are included in this study. This comprises all species known to be in cultivation here as well as a few native species that may occur with crop seeds. Some species may produce a variety of plant types, but the structural features of the seeds remain remarkably constant.

Sometimes seeds are milled rather severely to remove the cobwebby hairs on the lemmas. Allowance for the absence of hairs due to milling should always be taken into consideration in making an identification. To illustrate, some species produce a copious basal web, a tuft of long hairs arising at one point just above the callus. This basal web may be partially or completely removed in milling and such a seed could be confused with a species that normally does not develop a basal web. The illustrations show an undamaged floret at the left in each figure and, in addition, florets from all parts of a spikelet as they may appear after processing.

The gross diagnostic features may be recognized under the low magnification of 6 × or 10 ×. Finer details, such as the character of the pubescence, may require a magnification of 20 × or 40 ×.

Similar-Appearing Species and Varieties

P. juncifolia, alkali bluegrass, and *P. ampla*, big bluegrass:—Bulk samples and individual seeds of these two species are distinguishable if the seed has not been too severely milled. *P. ampla*, especially the variety Sherman big bluegrass, is in wider use than *juncifolia* at the present time.

P. glaucantha, glaucantha bluegrass:—The species is now in limited production. The plants have been confused with *P. compressa*, Canada bluegrass, which they resemble in having flattened culms. Seeds of *glaucantha*, as shown in the seed key, are quite similar to *interior* and *nemoralis*, but the three species appear to be distinguishable.

Poa pratensis, Kentucky bluegrass, and varieties:—The varieties of Kentucky bluegrass now in commerce are described and compared on page 19.

Key to *Poa*

The seed key is arranged in two sections: Section 1 (p. 18), the intermediate nerves of the lemma are evident to the base; section 2 (p.19), the intermediate nerves are lacking or evident only on upper half of the lemma.

Section 1

Intermediate nerves of lemma are sharply etched fine lines evident to the base; apex of lemma folded; basal web present except in *annua*.

1a. Pubescence on midnerve and marginal nerves only, the pubescence long and dense; lemma 5–6½ mm. long; hairs on keels of palea short and wide-spaced on upper half, long and dense on lower half (pl. XXIII, 83,D); apex of palea deeply V-notched .... *P. arachnifera*, Texas bluegrass (pl. XXII, 76).

1b. Pubescence on all nerves of lemma; length 2½–3 mm.

2a. Pubescence on nerves long and dense; lemma long-pointed, thin, tends to wrinkle; grain short and thick; hairs on keels of palea long and dense, do not extend to tip of palea (pl. XXIII, 83,E). *P. annua*, annual bluegrass (pl. XXII, 75).

2b. Pubescence on nerves of lemma short, confined to lower half; nerves very fine sharply etched lines.

3a. Lemma light to dark straw-color with a spot of darker brown at the base, the intermediate and marginal nerves usually not evident to tip of lemma; mass of seed evenly distributed, in lateral view both sides tend to be parallel or the dorsal side is slightly arched (see page 20); hairs on keels of palea coarse, wide-spaced, do not extend to tip of palea (pl. XXIII, 83,A) .... *P. pratensis*, Kentucky bluegrass (pl. XXII, 71).

3b. Lemma evenly brownish, thin and tight-fitting on the grain, the intermediate and marginal nerves usually evident to the tip; in lateral view the dorsal side tends to be strongly arched, the ventral side straight; palea with a sharp median crease, the hairs on the keels very minute and dense, extend to tip of palea (pl. XXIII, 83,B). *P. trivialis*, rough bluegrass (pl. XXII, 73).

SECTION 2

Intermediate nerves of lemma lacking or evident only on upper half of lemma, the nerves not sharply etched fine lines; basal web present or lacking; lemma light straw-color.

1a. Pubescence confined to nerves of lemma; basal web present.
2a. Hairs on keels of palea coarse, wide-spaced, do not extend to tip of palea (pl. XXIII, 83,A); mass of seed concentrated about midway (see page 20), in lateral view both ventral and dorsal sides taper to the base and apex from the thickened middle; apex of lemma obtuse and flared out; length 2½–3 mm.____*P. pratensis* var. Merion Kentucky bluegrass (pl. XXII, 72).
2b. Hairs on upper portion of keels of palea short, fine, dense, extend to tip of palea, becoming longer and wide-spaced below the middle (pl. XXIII, 83,C).
3a. Apex of lemma obtuse and flared out, the margins hyaline; length 2–2½ mm.
4a. Rachilla stout, with prominent apical disk; lemma light straw-color; width across palea ¾ mm.
P. compressa, Canada bluegrass (pl. XXII, 74).
4b. Rachilla slender, callus tends to be pointed; lemma narrow in proportion to length, very loose on the grain, the apex often gold-tipped or purplish; width across palea ½ mm.
P. palustris, fowl bluegrass (pl. XXIII, 79).
3b. Apex of lemma long-pointed, folded at the top; length 3–3½ mm.
5a. Rachilla pubescent, the hairs stiff and bristly.
P. nemoralis, wood bluegrass (pl. XXIII, 77).
5b. Rachilla not bristly pubescent.
6a. Rachilla glabrous, finely granular.
P. glaucantha, glaucantha bluegrass.
6b. Rachilla with short prickle-like hairs or coarsely granular.
P. interior, inland bluegrass (pl. XXIII, 78).
1b. Pubescence not confined to the nerves of lemma; basal web lacking; wide hyaline margins of lemma flared out at the top.
7a. Lemma 3–3½ mm. long; rachilla short, stout, with prominent apical disc; hairs on keels of palea long on lower half and very short above, extend to tip of palea (pl. XXIII, 83,F); pubescence on nerves long and dense, sparse and short between the nerves_______ *P. arida,* plains bluegrass (pl. XXIII, 80).
7b. Lemma 4–5 mm. long; rachilla slender, about ⅓ the length of the palea.
8a. Hairs on keels of palea short, dense, extend to tip of palea sparsely pubescent.
9a. Lemma coarsely granular and copiously short-hispid, the hairs prickle-like and rough to the touch, especially on the midnerve.
P. juncifolia, alkali bluegrass (pl. XXIII, 81).
9b. Lemma finely granular and sparsely short-pubescent, smoother than *juncifolia*__________________*P. ampla,* big bluegrass (pl. XXIII, 82).
8b. Hairs on keels of palea mostly long and wide-spaced, usually not extending to the tip of the palea; lemma and palea sparsely pubescent with short stiff hairs_______________________________*P. scabrella,* pine bluegrass.

Poa pratensis, Kentucky Bluegrass, and Varieties Compared

Several varieties of Kentucky bluegrass have come into production in recent years. Individual seeds of the variety Merion Kentucky bluegrass are quite readily identified, as shown in the seed key and illustrations. The varieties Arboretum and Troy are in limited use and differ from the species chiefly in the larger size of the seeds. Seeds of the variety Troy, a hay type, average 3½ to 4 mm. in length.

The varieties Park, Delta, and Newport are important lawn grasses. With practice it may be possible to identify seed of these varieties in the bulk. Individual seeds, because of natural variation, may be too similar to the larger northern-grown seed of common Kentucky bluegrass to be distinguishable with certainty.

Assuming that the reader is familiar with the diagnostic features of Kentucky bluegrass and the variety Merion Kentucky bluegrass, the following observations may be useful as a guide for the identification of bulk samples of the varieties Park, Delta, and Newport.

Inasmuch as the lemma is quite sharply keeled on the midnerve, the shape of the seed is determined from the lateral plane as the seed lies on its side.

The mass of the grain in certain varieties is evenly distributed so that the seed is approximately the same width throughout its length. The ventral and dorsal sides tend to be parallel, or the ventral side is straight and the dorsal side is slightly arched, as in common Kentucky bluegrass.

In other varieties the mass of the grain is concentrated about midway or lower and the ventral and dorsal sides may both taper to the apex and base from the thick median portion, as in the variety Merion Kentucky bluegrass. This shape difference may be less obvious on the terminal floret of a spikelet.

Color may vary somewhat in the different areas of production. The darker brown area at the base of the lemma, which is characteristic of some species, appears to be constant, differing only in degree.

Key to *Poa pratensis*, Kentucky Bluegrass, and Its Four Varieties: Merion, Park, Delta, Newport

1a. Intermediate nerves fine sharply etched lines evident to the base of the lemma; mass of seed evenly distributed; lemmas pointed, folded at the tip, not markedly wider than the palea.

2a. Color brownish, with a darker brown spot at the base; length 2¾ — 3 mm., width of lateral plane ¾ mm.
P. pratensis, Kentucky bluegrass (northern grown) (pl. XXII, 71).

2b. Color evenly brownish or occasionally darker at the base; length 3–3½ mm., width of lateral plane ¾ mm.
P. pratensis, Kentucky bluegrass var. Park.

1b. Intermediate nerves lacking, or obscure and wrinkle-like (not fine sharply etched lines) and evident only part way of the lemma.

3a. Mass of seed evenly distributed; lemmas not markedly wider than the palea, pointed and folded at the tip; pale straw-color with a darker area at the base; length 2½–3 mm., width of lateral plane about ½ mm.
P. pratensis, Kentucky bluegrass var. Delta.

3b. Mass of seed concentrated about midway or lower; lemmas with wide hyaline margins toward the top, much wider than the palea, the apex flared out.

4a. Intermediate nerves, if present, extend from the top to about the middle of lemma; pale straw-color, not darker at the base; hairs on keels of palea coarse, wide-spaced; length 2½–2¾ mm., width of lateral plane ¾–1 mm.----*P. pratensis*, Kentucky bluegrass var. Merion (pl. XXII, 72).

4b. Intermediate nerves, if present, extend from the base to about the middle of lemma; color brownish, darker at the base; hairs on keels of palea long and pointed, wide-spaced; length 3 mm., width of lateral plane 1 mm.----------------*P. pratensis*, Kentucky bluegrass var. Newport.

PUCCINELLIA AIROIDES, NUTTALL ALKALIGRASS

Ten species of *Puccinellia* are reported for the United States, but only one species is of agricultural interest. In addition to its value as a native forage grass, a form of this species is in limited cultivation under the name of Zawadke alkaligrass.

Florets are 2½–3 mm. long, ½ mm. wide. Lemmas are awnless and have hyaline margins that are markedly wider than the palea in the upper half, the nerves are mostly lacking.

TRIDENS FLAVUS, PURLETOP (PL. XXX, 202)

Florets are 3½–4 mm. long, often purplish. Lemmas are 3-nerved, the nerves exserted at the obtuse apex and densely pubescent on the lower half and on the ends of the callus. The broad palea is 3-lobed at the apex and bowed out below, the hairs on the keels very short, fine and dense, the rachilla slender and glabrous. The species may occur as a contaminant in agricultural seeds.

2. Hordeae—Barley Tribe

The tribe *Hordeae* includes valuable forage and range grasses and the important cereals, barley, wheat, and rye. Included also are some very objectionable weed species such as quackgrass, *Agropyron repens*, and goatgrass, *Aegilops* spp. Seeds of the cereals are so well known they are not included in the descriptions that follow. The main problem in cereals concerns the identification of horticultural varieties, which is not a part of this study.

The inflorescence of the plants in this tribe is a single terminal spike, the spikelets being in rows on opposite sides of a jointed or continuous rachis. The rachis may be variously modified. In some genera it is flattened or concave next to the spikelets as in *Lolium*, ryegrass; in other genera it is thickened and hollowed out, the spikelet being more or less embedded in the hollow as in *Aegilops*, goatgrass. The spikelet in certain genera consists of several florets as in *Lolium*, in other genera it is 1-flowered as in the wild-barleys, *Hordeum*.

Key to the Genera in the Barley Tribe

1a. Rachis separating into joints.
  2a. Rachis segment flat, sharp-edged, bears a cluster of three 1-flowered spikelets ---------- *Hordeum* (p. 26).
  2b. Rachis segment thickened above, hollowed out below and partly enclosing a 2-flowered spikelet ---------- *Aegilops* (p. 21).
1b. Rachis not separating into joints.
  3a. Rachilla segments flat, the sides parallel; florets oblong, scarcely narrowed at the base ---------- *Lolium* (p. 27). *Agropyron* (in part) (p. 22).
  3b. Rachilla segments not as above; florets taper to a narrow base.
    4a. Rachillas mostly thick, the sides more or less divergent toward the top. *Agropyron* (p. 22). *Elymus* (p. 22).
    4b. Rachilla of fertile spikelet long and threadlike ---------- *Hordeum* (p. 26).

AEGILOPS, GOATGRASS

Three species of *Aegilops* are listed for the United States, all of which are objectionable weeds in grain fields and range lands. The seed unit found as a contaminant with crop seeds is usually a single joint of the spike. The spikelets are placed edgewise at each node of the joint and fit closely into a groove in the rachis.

The joint consists of a thick and hardened segment of the rachis and a 2-flowered spikelet enclosed by a pair of hardened and rigid-awned glumes. Figure 4(*B*), page 8, shows a joint of a spike of *Aegilops cylindrica* and a detail of a glume.

Lemmas and paleas of the fertile florets are thinner than the glumes and somewhat translucent, the lemmas bearing three rigid awns of various lengths. The caryopsis resembles a long grain of wheat.

Key to *Aegilops*

1a. Glumes with one awn and a pronounced indentation at one side of the awn, the awn scabrous; glumes glabrous or nearly so; length of joint 10–12 mm. *A. cylindrica*, jointed goatgrass (fig. 4(*B*), p. 8).
1b. Glumes with more than one awn, the glumes and awns very scabrous.
2a. Glumes with three strong awns, 4–8 cm. long; length of joint 10 mm. *A. triuncialis*, barb goatgrass.
2b. Glumes with four stiff and spreading awns, 2–3 cm. long; length of joint 6–7 mm. *A. ovata*, ovate goatgrass.

AGROPYRON, WHEATGRASS; *ELYMUS*, WILD-RYE

The structural features of the florets in these two genera are very similar and the species under consideration are treated together in this study. This includes all species known to be in cultivation in this country or that may occur as incidental weeds in crop seeds: *Agropyron*, 14 species and several varieties; and *Elymus*, 7 species and 1 variety. (See plates I–IV and XXVI.) The seed unit in commercial seed is the mature floret. Occasionally the glumes persist on the basal florets, as in *E. virginicus*, Virginia wild-rye.

The several-flowered spikelets of *Agropyron* are sessile and solitary at each node of the rachis, the spikelets placed flatwise against the rachis. The two to several-flowered spikelets of *Elymus* are sessile and usually in pairs at each node of the rachis. The rachilla is distorted at the base so that the spikelets lie somewhat edgewise to the rachis. As a result of this slight twisting, the callus of the basal florets may be slightly off-center. This is especially common in *E. virginicus*.

New horticultural varieties of these grasses are being developed from time to time and appear in commerce. Some of these are included in the seed key. As a rule, seeds of the named varieties are quite uniform in structural features and lack the range of variation that is often found in unselected lots of a species.

The seeds in the *Agropyron-Elymus* group are fairly large so it is usually possible to see the diagnostic features clearly with a 6 × hand lens. Seeds that have been rubbed or damaged in processing may need to be examined under a magnification of about 20 ×.

The length of florets (exclusive of awns) usually varies in a spikelet from the basal floret to the terminal floret, and the measurements stated in the seed key should be considered as approximate only.

As shown in the illustrations, the length of the rachilla segment varies with its position in the spikelet and may range from very short on the basal floret to long and slender on the terminal floret. The rachilla segment on the terminal floret usually does not develop any diagnostic character.

Hulled caryopses are sometimes present in a seed sample and it is important to distinguish the caryopses of the objectionable quackgrass (*A. repens*) from those of the similar-appearing crop species.

Similar-Appearing Species

Bulk samples of *A. cristatum*, fairway crested wheatgrass, are distinguishable from *A. desertorum*, standard crested wheatgrass, by the markedly smaller size of the seeds. But because of natural variation within a species, individual seeds of the two species can not be identified with certainty. Seeds of *A. desertorum* and *A. sibiricum*, Siberian wheatgrass, are so similar that bulk samples and individual seeds may be indistinguishable.

Undamaged seeds of *A. subsecundum*, bearded wheatgrass, are distinguishable from *A. trachycaulum*, slender wheatgrass, but they are not always distinguishable when the awn and lemma tips are broken off. *A. trachycaulum* comprises a somewhat variable group of plants which is segregated into varieties and forms by some authors. The florets may be awnless, short-awned, or fairly long-awned. The seeds in commerce are usually as described in the seed key.

A long narrow grain is characteristic of *A. inerme*, beardless wheatgrass, a feature that is apparent even when the tip of the lemma is broken off. Seeds of *A. inerme* and *E. glaucus*, blue wild-rye, that have been damaged in processing may appear to be similar and the features listed in the seed key (sec. I, *5a*, and sec. 2, *4a*) should be carefully compared. Seeds of *A. spicatum*, bluebunch wheatgrass, may be indistinguishable from *A. inerme* if the awn is not present.

It is not known whether there is intergrading in seed characters between *A. intermedium*, intermediate wheatgrass, and the closely related *A. trichophorum*, hairy intermediate wheatgrass. It is quite possible that some seeds may be indistinguishable.

Some authors report that plants of *A. dasystachyum*, thickspike wheatgrass, and *A. riparium*, streambank wheatgrass, tend to intergrade. An examination of the available material indicates that seeds of the two species are distinguishable. Further investigation is needed.

Seeds of the giant wild-ryes, *Elymus cinereus* and *E. condensatus* are very similar. *E. cinereus* appears to be the more widely distributed of the two species and it is more likely to occur as a field weed.

The following species that normally have glabrous lemmas may produce a variable proportion with pubescent lemmas: *A. desertorum*, *A. intermedium*, *A. smithii*, and *E. virginicus*. These species will appear in both categories in the seed key.

Key to *Agropyron* and *Elymus*

The seed key is arranged in two sections:

Section 1. Lemmas essentially glabrous (p. 23).

Section 2. Lemmas scabrous, hispid or villous, at least in part (p. 25).

Section 1

Lemmas essentially smooth and glabrous or sparsely minutely pubescent across the tip in *A. trachycaulum* and *A. subsecundum*.

1a. Lemmas compressed laterally throughout; palea concavity deeply V-shaped, the apex folded, hairs on the keels wide-spaced, short and stout.

2a. Length 6–7 mm., width 1 mm., awns various, sometimes lacking.
A. desertorum, standard crested wheatgrass (pl. III, 9).
A. sibiricum, Siberian wheatgrass.

2b. Length 5–6 mm., width ¾ mm.; all lemmas awned, awns about half the length of the lemma.
A. cristatum, Fairway crested wheatgrass (pl. III, 10).

1b. Lemmas rounded or flattened on the back, or if compressed laterally only slightly so, the palea well exposed; length 8–10 mm., width 1¼–2 mm., hairs on keels various.

3a. Rachilla villous, callus hairs continuous across the back; lemma sparsely pubescent across the tip, the hairs very short and fine.

4a. Apex of lemma bifid; awn about twice the length of lemma; palea dull and scurfy, markedly shorter than the lemma, the keel nerves commonly exserted at the obtuse apex.
A. subsecundum, bearded wheatgrass (pl. I, 2).

4b. Apex of lemma not bifid; awnless or with short awn; palea glabrous and lustrous, sometimes minutely pubescent on upper portion, tapers to a fine V-notch at the apex---*A. trachycaulum*, slender wheatgrass (pl. I, 1).

3b. Rachilla glabrous or very short-pubescent (some exceptions in *Elymus triticoides*); callus hairs short, confined to ends of callus or lacking.

5a. Lemma linear-oblong with several fine nerves at the top, length 10–12 mm., width 1¼–1½ mm.; palea deeply concave, glabrous and lustrous, hairs on keels short, close-spaced; rachilla long, slender (except on basal florets), with a fine brown line on the rim of the turgid apex; callus wide (vertically), lustrous.

6a. Lemma awnless or with short, weak awn.
A. inerme, beardless wheatgrass (pl. I, 3).

6b. Lemma with stout, strongly divergent awn 1–2 cm. long.
A. spicatum, bluebunch wheatgrass. (pl. I, 4).

5b. Lemmas other than linear-oblong, obscurely 3- 5-nerved at the top; rachillas mostly stout, the sides more or less divergent except in *A. repens*.

7a. Lemma elliptic, tapers to a narrow base, length 8–9 mm., width 1½–2 mm.; callus wide (vertically), glabrous, lustrous; glumes commonly persisting on the basal florets.

8a. Lemma yellowish brown or tawny, glazed and semi-glossy, awns short and weak, margins of lemma extend to keels of palea or nearly so.

Rachilla sparingly pubescent, occasional hairs fairly long; palea glabrous or sparsely pubescent, scarcely concave, with a median groove or ridge below--------*E. triticoides*, beardless wild-rye.

8b. Lemma pale, with yellowish tinge, coarsely granular and dull, sometimes sparsely scabrous across the top (see sec. 2, *11a*); narrows abruptly to a long, narrow base; awns straight, up to 10 mm. or longer.

Rachilla glabrous; palea broad, smoothly concave, glabrous or minutely pubescent at the top, the apex obtuse or with a shallow notch, the keels often inturned and flattened down in lower portion.
E. virginicus, Virginia wild-rye (pl. XXVI, 143).

7b. Lemmas lance-shaped or oblong, scarcely narrowed at the base, pale straw-color, or with brownish or greenish tinge in *A. repens*.

9a. Lemma slender, long pointed, length 8–9 mm., width 1¼–1¾ mm.; awns ¼–⅓ the length of the lemma; palea slender.

10a. Sinus U-shaped, the base of rachilla fully exposed; rachilla sides not divergent; lemma smooth, slightly glazed and lustrous.

Palea smoothly concave or with a longitudinal fold in lower half, glabrous or sparsely pubescent, not folded at the top, the apex truncate, obtuse or with a broad, shallow notch (pl. III, 11A); hairs on keels of palea usually short with a broad base.
A. repens, quackgrass (pl. II, 7).

10b. Sinus V-shaped, the base of rachilla partially obscured by the margins of lemma, occasionally narrowly U-shaped; rachilla sides strongly divergent; lemma coarsely granular, dull.

Palea smoothly concave or grooved in lower half, often folded at the tip, the apex deeply V-notched (pl. III, 11B), the surface scurfy and sparsely pubescent; hairs on keels of palea usually longer and finer than in *repens* (pl. II, 7). (See also sec. 2, *10b*.)
A. smithii, western wheatgrass (pl. II, 5).

9b. Lemmas broad, obtusely pointed, awn tipped or short-awned, length 8–10 or 12 mm., width 1¾–2 mm.; paleas broad and obtuse.

11a. Palea concavity broad and shallow or with only a slight crease toward the base, dull, scurfy and sparsely pubescent; hairs on keels of palea long, fine, close-spaced in upper portion.

Rachilla short (about ⅕ the length of the palea), thick, the sides strongly divergent; lemma dull, markedly exceeding the tip and sides of the palea in upper ⅓; grain relatively thin with a broad, shallow concavity on the ventral side.
A. elongatum, tall wheatgrass (pl. IV, 13).

11b. Palea concavity deeply V-shaped, at least in lower portion, lustrous and translucent, glabrous or minutely pubescent; hairs on keels of palea not long and close-spaced as in *elongatum*.
Rachilla longer and more slender (about ⅓ the length of the palea), the sides not strongly divergent; lemma lustrous, scarcely exceeding the tip and sides of the palea at the top; grain thick, with a deep V-shaped crease on the ventral side. (For the pubescent form, see sec. 2, *6a*.)
A. intermedium, intermediate wheatgrass (pl. IV, 12).

Section 2

Lemmas scabrous, hispid or pubescent, at least in part.

1a. Lemmas compressed laterally throughout, sparsely pubescent, the hairs long and lax; length 6–7 mm.; otherwise as in section 1, *1a*.
A. desertorum, standard crested wheatgrass (pl. III, 9).
A. sibiricum, Siberian wheatgrass.

1b. Lemmas rounded or flattened on the back, or if keeled only in part; length 7–10 or 12 mm., width 1–1½ mm.

2a. Paleas almost obscured by the strongly inrolled, but not flattened-down keels; lemma very coarsely granular.

3a. Lemma linear-oblong, stiff, minutely scabrous and sparsely hispid across the top and along the sides; length 7–8 mm., width 1 mm.; awn coarse, straight or flexuous, 20–30 mm. long.
Palea concavity narrow, deeply V-shaped, flattened out at the narrow, notched tip, hairs on keels short, stubby, wide-spaced; rachilla thread-like; callus wide (vertically), glossy, entire callus area evident in ventral view........................ *E. caput-medusae*, medusa-head wild-rye.

3b. Lemma lance-shaped, awnless or with short, weak awn, soft, and papery in texture, scabrous and usually copiously pubescent with bristly hairs, the hyaline margins extend to keels of palea or nearly so; length 8–10 mm., width 1–1½ mm.
Palea concave, has sunken appearance, densely pubescent and scurfy, hairs on keels fine, wide-spaced; rachilla stouter, finely pubescent; callus inconspicuous, lies in vertical position, the callus area not evident in ventral view... *E. junceus*, Russian wild-rye (pl. XXVI, 142).

2b. Paleas exposed throughout.

4a. Lemma bifid, linear-oblong, sometimes purplish, coarsely granular and sparsely scabrous, especially on upper portion, length 8–10 mm., width 1¼–1½ mm.; awn weak, 6–8 mm. long.
Palea deeply concave, dull, pubescent in upper portion, hairs on keels short, wide-spaced: rachilla long, slender, about ⅓ the length of the palea; callus wide (vertically), dull. (Compare with sec. 1, *5a*, p. 24.)
E. glaucus, blue wild-rye (pl. XXVI, 141).

4b. Lemma not bifid or linear-oblong.

5a. Lemma broad, awn-tipped or short-awned, glabrous or with long lax hairs across the top and along the sides, the grain thick; length 9–10 mm., width 1¾–2 mm.

6a. Lemma finely granular, slightly glazed and lustrous on lower half; other features as in section 1, *11b*, page 25.
A. intermedium, intermediate wheatgrass (in part) (pl. IV, 12).

6b. Lemma coarsely granular, dull, glabrous on lower half; other features as in section 1, *11b*, page 25.
A. trichophorum, hairy intermediate wheatgrass (pl. IV, 14).

5b. Lemma slender, the pubescence various, not confined to the top and sides, the grain relatively thin; length 8–10 or 12 mm., width 1½ mm.

7a. Lemma awnless or with short awn.

8a. Apex of palea truncate or with shallow notch or indentation, the keels mostly inturned and flattened down in lower half; rachilla villous; basal hairs continuous across the callus or confined to the ends.

9a. Lemma thin and papery, villous or short-pubescent; palea sparsely pubescent.
A. dasystachyum, thickspike wheatgrass (pl. IV, 15).

9b. Lemma not thin and papery, sparsely hispid or slightly rough to the touch and villous on the sides near the base; palea minutely pubescent.. *A. riparium*, streambank wheatgrass (pl. II, 8).

8b. Apex of palea with an extremely deep V-notch.
10a. Rachilla slender, villous, commonly with a fine brown line on the rim of the turgid apex; lemma sparsely hispid and villous along the sides and at the base, sometimes purplish.
E. cinereus, giant wild-rye.
10b. Rachilla thick, the sides strongly divergent, short-pubescent; lemma coarsely granular and sparsely short-hispid; otherwise as in section 1, *10b*.
A. smithii var. *molle*, hairy western wheat grass (pl. II, 6).
7b. Lemma long-awned (10–30 mm.), the awns scabrous; keels of palea mostly inturned and flattened down to near the top.
11a. Awns straight; lemma narrowed abruptly to a long and narrow base, scabrous or hirsute in certain varieties; otherwise as in section 1, *8b* ---- *E. virginicus*, Virginia wild-rye (pl. XXVI, 143).
11b. Awns long and divergent; lemma tapers evenly to a narrow base, dull, coarsely granular and sparsely long-pubescent, especially along the sides; palea sparsely pubescent, tapers to a narrow, notched apex; rachilla almost glabrous; callus wide (vertically), dull, glabrous or short-pubescent on the ends.
E. canadensis, Canada wild-rye (pl. XXVI, 140).

HORDEUM, WILD-BARLEY

Five of the more widely distributed weed species of *Hordeum* are described. The well-known cultivated barley *H. vulgare* is not included.

The jointed rachis of the spikes is flattened and sharp-edged and separates at maturity into individual joints, each joint bearing a cluster of three 1-flowered spikelets. The attachment of the spikelets to the rachis segment differs from most grasses in that the cluster of three spikelets remains attached at the top of the segment, and the naked lower portion of the segment appears like a stalk. These sharp-edged rachis segments, when mature, may be injurious to livestock.

The floret of the central spikelet of the triad is fertile and, in most species sessile, the rachilla prolonged as a long slender bristle nearly ¾ the length of the palea. The two lateral spikelets of the triad are sterile and pediceled, the lemma in some species being well-developed, in others scarcely expanded. The glumes are rigid and may be reduced to long scabrous bristles in certain species, in other species they are flattened and expanded.

The bristles are more or less broken in harvesting so the actual length is not usually evident in processed seed. The overall length of the rachis joint, which includes the cluster of three spikelets with its bristles and rachis segment, is included in the seed key only for the purpose of indicating the relative sizes of the different species.

The seed unit found in crop seeds may be an entire joint, consisting of the triad of spikelets with the attached rachis segment, or it may be only the fertile central floret with no accessory structures.

Structural Features of the Fertile Floret

The identification of a fertile floret when separated from the triad may be determined by the character of the palea and the shape of the floret.

In most grasses the width of the grain coincides with the area between the keel nerves of the palea. In some species of *Hordeum* the grain exceeds the keels of the palea to a marked degree, giving the keel nerves the appearance of being close together. Depending on the shape of the grain, the palea may be relatively flat and well exposed, or the sides of the grain may be inrolled forming two thick longitudinal rolls which almost touch down the middle and more or less completely obscure the palea.

Shape of the floret in outline is also a helpful feature. The florets of certain species are narrowly elliptic with both ends long-pointed. Other species are lance-shaped with a broad base and long-pointed apex.

The fertile, sessile florets of *jubatum* and *brachyantherum* are very similar. There may be intermediate forms between the two species. Florets of *brachyantherum* tend to average longer than *jubatum*.

Key to *Hordeum*

1a. Lemmas of sterile florets not expanded, all glumes bristle-like, or in *hystrix* one glume of each sterile spikelet and the sterile lemmas slightly expanded; fertile spikelets sessile, sterile spikelets pedicellate.

2a. Entire palea of fertile floret exposed or nearly so, flat, broadest about midway, usually with a prominent fold between the parallel keel nerves; fertile floret 5–8 mm. long, 1½ mm. wide; overall length of rachis joint 20 mm. or more, the bristles delicate and flexuous.
H. jubatum, foxtail barley (pl. XXVII, 160).

2b. Entire palea not exposed.

3a. A narrow strip of palea exposed between the sharply inturned keel nerves; fertile floret 7–10 mm. long, ½ mm. wide, the lemma sometimes purplish; overall length of rachis joint 14–15 mm., the bristles stiff.
H. brachyantherum (*H. nodosum*), meadow barley (pl. XXVII, 159).

3b. Entire palea obscured by the two thick rolls of the strongly inrolled caryopsis and palea; fertile floret 7–8 mm. long, 1 mm. wide; overall length of rachis joint 25 mm.
H. hystrix (including *H. gussonianum*), Mediterranean barley.

1b. Lemmas of sterile florets well-developed; not all glumes bristle-like.

4a. Fertile spikelet sessile, sterile spikelets pedicellate, the pedicels short and stout; first glume of sterile spikelets and both glumes of fertile spikelet expanded at the base.
Entire palea exposed or nearly so, flat, broadest about midway; fertile floret 5–6 mm. long, 1½ mm. wide; overall length of rachis joint 10–14 mm.
H. pusillum, little barley (pl. XXVII, 158).

4b. Both fertile and sterile spikelets pedicellate, the pedicels long and slender; glumes of fertile spikelet expanded, the margins ciliate; one glume of each sterile spikelet expanded and ciliate, the other a long bristle.
Palea completely obscured by the inrolled keels of the palea; fertile floret elliptic, the apex and base equally pointed, length 8–9 mm., width 1½ mm.; overall length of rachis joint 45–50 mm.
H. leporinum, wall barley (pl. XXVII, 161).
(*H. murinum* of American authors.)

LOLIUM, RYEGRASS

The several-flowered spikelets of *Lolium* are placed edgewise in alternate shallow cavities on a continuous rachis, the first glume lacking except on the terminal spikelet. Unlike most grasses, the rachilla segment on the basal floret of a spikelet is extremely long and stout. The seed unit in commercial seed is the mature floret, the free caryopses do not commonly occur.

The genus is represented here by two important crops: *L. perenne*, perennial ryegrass, and *L. multiflorum*, Italian ryegrass. *L. multiflorum* var. Westernwolth and *L. rigidum* (or *L. subulatum*), Wimmera ryegrass, are sometimes grown experimentally. The seeds appear to be indistinguishable from *L. perenne* and *L. multiflorum*.

The terms "common" ryegrass and "domestic" ryegrass are sometimes used in the trade. Neither term is recognized as the name of a kind or variety. "Common" ryegrass is interpreted to be synonymous with "ryegrass." Thus, seed sometimes designated as "common" may be either perennial or Italian ryegrass or a mixture of the two kinds. "Domestic" ryegrass is interpreted to indicate the origin of the seed.

Three weed species may occur as contaminants with crop seeds: *L. temulentum*, darnel; *L. persicum*, Persian ryegrass; and *L. remotum*, slender darnel.

Awns are usually broken off in processing so this feature is not included in the seed key. All florets of *L. multiflorum* are awned or only rarely awnless. *L. perenne* is awnless or occasionally with a short, weak awn. *L. persicum* is long-awned. *L. temelentum* and *L. remotum* have both awned and awnless forms, the awns of *remotum* being short and weak.

The comparative size and shape of these four species is shown diagrammatically in figure 8. For a comparison of the seeds of ryegrass and the somewhat similar seeds of the tall fescues, refer to pages 16-17.

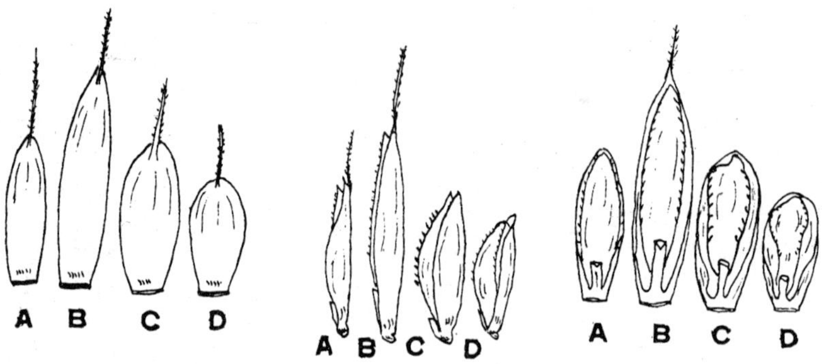

FIGURE 8.—Comparative size and shape of four species of *Lolium*, shown in ventral, lateral, and dorsal views: (A) *L. multiflorum*, (B) *L. persicum*, (C) *L. temulentum*, (D) *L. remotum*.

Key to *Lolium*

1a. Seeds oblong or lance-shaped, scarcely narrowed at the base, flattened dorso-ventrally, about the same thickness throughout.

2a. Lemma short-pointed, 6–7 mm. long, 1¼–1½ mm. wide; palea short-pointed, hairs on keels short and blunt.
L. perenne, perennial ryegrass (pl. XVIII, 50).
L. multiflorum, Italian ryegrass (pl. XIX, 51).

2b. Lemma long-pointed, 8–10 mm. long, 1¾–2 mm. wide; palea tapers to a long and narrow apex, hairs on the keels fairly dense and pointed.
L. persicum, Persian ryegrass (pl. XVIII, 49).

1b. Seeds oval, not flattened dorso-ventrally; in lateral view the dorsal side straight, ventral side strongly arched.

3a. Seed thickest about midway; palea broad, narrowed near the top to a short-pointed apex, usually with a transverse wrinkle above the tip of the grain; hairs on keels of palea short and blunt; length 6–7 mm., width 3–3½ mm. .... *L. temulentum*, darnel (pl. XIX, 52).

3b. Seed thickest in upper half; palea broad and flared out at the top, hairs on keels of palea inconspicuous, short, fine and dense; length 4–4½ mm., width 1–1½ mm. .... *L. remotum*, slender darnel (fig. 8, p. 28).

3. Aveneae—Oat Tribe

The inflorescence and spikelets are similar to those of the fescue tribe (fig. 4, p. 8), but with much enlarged glumes, usually covering the entire spikelet. Spikelets are mostly 2-flowered, the lemmas are rounded on the back, thin and chaffy or stiffer, or compressed laterally in *Holcus*. The seed unit may be a single floret or a spikelet.

Key to the Genera of *Aveneae*

1a. Paleas hyaline, completely exposed, tend to spread away from the lemma.
  2a. Awns weak or lacking.
    3a. Rachillas long and densely pubescent, the hairs long and spreading. *Trisetum* (p. 34).
    3b. Rachillas sparsely long-pubescent.
      4a. Lemmas coarsely granular, pubescent with short appressed hairs. *Koeleria* (p. 34).
      4b. Lemmas finely granular, essentially glabrous..*Sphenopholus* (p. 34).
  2b. Awns on sterile florets stout, geniculate, twisted and spirally striped tan and brown; rachillas on fertile florets threadlike..*Arrhenatherum* (p. 30).
1b. Paleas not hyaline, do not spread away from the lemma.
  5a. Spikelets and florets strongly compressed laterally; glumes equal, longer than the spikelet; fertile lemmas smooth, stiff and glossy..*Holcus* (p. 33).
  5b. Spikelets and florets not compressed laterally.
    6a. Lemmas dark brown, becoming hyaline and whitish toward the top, with whitish hispid hairs across the upper portion and along the sides; length 1½–3 mm., width ¼–⅓ mm....................*Aira* (p. 29).
    6b. Lemmas not as above.
      7a. Florets with a distinct cavity at the point of attachment, or with a piece of rachilla segment persisting below............*Avena* (p. 30).
      7b. Florets without a distinct cavity at the base.
        8a. Lemmas pointed; rachilla and callus copiously long-pubescent, the hairs whitish, spreading.
          9a. Lemmas pale, lustrous, with a weak awn from near the base or awnless; ventral side bulged out; length 2½ mm. *Deschampsia* (in part) (p. 33).
          9b. Lemmas golden yellow, dull, with a geniculate, twisted awn from near the base; ventral side flat; length 4 mm. *Deschampsia* (in part) (p. 33).
        8b. Lemmas flared out toward the top, the apex deeply bifid, a stout twisted and geniculate awn arising between the acute hyaline-edged lobes; paleas broad, the apex truncate; length 3½–5 mm. *Danthonia* (p. 33).

AIRA, HAIRGRASS

The 2-flowered spikelets of *Aira* disarticulate above the glumes and between the florets. Lemmas of florets are brown, becoming whitish and hyaline toward the top, with whitish hispid hairs across the top and along the sides. Apex of lemmas terminates in two long slender teeth about ¾ mm. long, the callus densely pubescent with whitish spreading hairs. Paleas are sunken below and partially obscured by the inrolled margins of the lemma. Rachilla segments are extremely short and present only on the basal florets.

Two species may occur as incidental weeds in crop seeds. The seed unit in both species is a single floret.

Key to *AIRA*

1a. Florets 2½–3 mm. long, ½ mm. wide; lemmas of all florets with geniculate, twisted awns, about 4 mm. long, arising below the middle of the lemma. *Aira caryophyllea*, silver hairgrass (pl. XXIV, 84).

600705 O—62——3

1b. Florets 1½–1¾ mm. long, ½–⅓ mm. wide; lemma of basal floret awnless or with a short weak awn, the upper floret with geniculate twisted awn about 3 mm. long.----*Aira elegans* (*A. capillaris*), fine hairgrass (pl. XXIV, 85).

ARRHENATHERUM, OATGRASS

One species of *Arrhenatherum* is produced as a crop in this country. Spikelets are 2-flowered and disarticulate above the glumes, the lower floret is staminate and the upper floret is perfect. The lower, staminate floret remains attached to the upper fertile floret at maturity. The seed unit may be the entire spikelet or a hulled caryopsis.

Arrhenatherum elatius, tall oatgrass (pl. XXIV, 103).—Lemmas are 8–10 mm. long, 5- to 7-nerved, minutely scabrous across the top and with scattered long hairs down the middle. Callus hairs are long and copious. Lemma of the upper floret bears a short straight awn just below the tip. The lower staminate floret bears from near the base a long geniculate awn, twisted and spirally striped with brown and tan. The fertile floret disarticulates by fracture and the rachilla is extended as a short bristle.

Paleas are hyaline and narrow, the keel nerves close together. The paleas tend to spread away from the lemmas.

Hulled caryopses are approximately 4–5 mm. long and nearly cylindrical in shape. The color is light yellowish or pale brownish.

Arrhenatherum elatius var. Tualatin is an improved variety of tall oatgrass. Florets and caryopses are doubtfully distinguishable from those of the species.

AVENA, OATS

Spikelets in the genus *Avena* may be 2- or 3-flowered, with elongated pointed glumes. The mature spikelet disarticulates above the glumes and between the florets in most species and the type of abscission is an important feature in distinguishing the species. The seed unit in commercial seed is the mature floret or, infrequently, the hulled caryopsis.

The primary or basal floret in a spikelet may possess some features that are entirely lacking in the upper florets. It is therefore important to be familiar with the morphology of all florets in the spikelet of any given species.

The kinds under consideration include the cultivated oats (*A. sativa* and *A. byzantina*) together with the fatuoid or false wild forms frequently occurring with them, and certain wild species that have become widely distributed and are utilized for hay or occur as weeds in cultivated fields (*A. fatua, A. barbata, A. strigosa*). Included also is *A. sterilis* from which, according to some authors, *A. byzantina* is believed to have been derived.

Cultivated oats.—Cultivated oats belong to one of two species or derivatives thereof: *Avena byzantina*, the red oat group, or *A. sativa*, the northern oat. The red oat is represented in the illustration by the varieties Fulghum and Red Rustproof, and the northern oat by the variety Victory. It is not always possible to determine from the character of the seed whether a variety belongs to the *byzantina* or the *sativa* group.

The mature spikelets break apart by fracture. It will be observed that the basal florets of the *byzantina* varieties have prominent basal cavities lying in an oblique position with reference to the long axis of the seeds. The cavities of the basal florets in *sativa* are smaller and lie in a vertical position. The rim of the cavity in both species is thin and slightly uneven. The upper floret usually retains a splintered portion of the rachilla segment below it. This is in contrast to the thick smooth rim of the callus on all florets of *A. fatua*, wild oat, and the homozygous fatuoids.

Fatuoids.—Three explanations have been advanced in explanation of the origin of fatuoids: Natural hybridization, chromosome aberration, and gene mutation. The mutation theory appears to be the most widely accepted. Fatuoids occur in both northern and southern oat varieties, possibly with greater frequency in the latter. They tend to resemble in size, color, and texture the variety of oats with which they occur. There is perhaps one exception, an occasional black fatuoid may appear in red oats.

The homozygous fatuoid bears a superficial resemblance to *A. fatua* but, as will be shown, the two are readily distinguished. The heterozygous fatuoid bears little or no resemblance to *A. fatua*. The upper florets of the spikelet and some of the awnless basal florets are doubtfully distinguishable from the cultivated oats with which they occur.

Avena fatua, wild oats (pl. XIII, 33).—The florets of wild oats may vary considerably in size, color, and character of pubescence. The color is usually gray, brownish or black, occasionally yellowish. Lemmas may be densely pubescent or almost glabrous and the hairs around the basal cavity may be long and spreading or short and stubby. Some authors divide *A. fatua* into several subspecies and varieties but for practical purposes these are considered here as one species.

Diagnostic Features

Types of abscission.—There are three types of articulation of the spikelet from its pedicel and of the florets from each other: (1) Complete abscission; (2) partial abscission; and (3) fracture. The different types of abscission are shown in plate X.

Complete abscission.—A thick layer is developed at the base of the floret and the mature floret breaks off with a smooth, well-developed callus around a prominent basal cavity.

Partial abscission.—The thickened layer is not developed around the entire basal cavity of the floret, usually only on the sides. A small portion on the front and at the back remains lightly solidified and the floret does not break off smoothly at these points. This type of abscission is found commonly in the heterozygous fatuoid.

Fracture.—The points of articulation have become lightly solidified around the entire basal cavity. The floret may break off more or less regularly along the line where the abscission layer would normally develop but there is no thick callus as in type (1), or the rachilla segment may break or splinter at some point. The basal cavity is often much reduced in size. This type of abscission is found in cultivated oats.

Caryopses of *Avena*

The character of the caryopsis is particularly helpful in distinguishing *A. fatua* and the fatuoids when other diagnostic features of the floret are in doubt. Shape of the floret in outline is not entirely dependable as some degree of overlapping may occur, especially in terminal florets. The character of the ventral side of the caryopsis and the appearance of the scutellum on the dorsal side appear to be fairly constant and may serve as reliable features in distinguishing the two forms. The ventral side of *A. fatua* tends to be flat with a fine groove down the middle. The scutellum is not evident on the dorsal side or it appears as a faint whitish line. In the fatuoids the ventral side is transversely arched by two thick longitudinal folds with a deep groove between them; the scutellum is triangular in shape, large and distinct. (pl. XIV, 37A-F).

Key to Wild and Cultivated Species of *Avena* and the Fatuoids

1a. All florets of spikelet disarticulate by complete abscission; all florets with geniculate, twisted awns, the bend of the awn well above the tip of the lemma or equal to it.

2a. Lemma terminates in two minute teeth; rachilla segment copiously hairy, the hairs long and stiff; lemma sparsely long-hairy or glabrous; basal hairs long and spreading or short and stubby.

3a. Florets broadly elliptic, the upper portion of palea exposed; in edge view markedly thickened at the base of awn; caryopsis markedly thickened in lower half, the scutellum large and distinct (pl. XIV, 37, *D*) -- Fatuoid, homozygous (pl. XII, 31).

3b. Florets narrowly elliptic, the upper portion of palea obscured by the inrolled lemma; in edge view only slightly thicker at the base of awn; caryopsis slightly thickened about midway, the scutellum not evident or only a faint whitish line (pl. XIV, 37,C) -- *A. fatua*, wild oat (pl. XIII, 33).

2b. Lemma linear, terminates in two hyaline-edged bristles; lemma, rachilla segment and base copiously hairy, the hairs long, stiff, and spreading; in edge view not evidentally thickened at the middle; caryopsis slender, the scutellum distinct (pl. XIV, 37, *F*) -- *A. barbata*, slender oat (pl. XIII, 35).

1b. All florets of spikelet not disarticulating by complete abscission.

4a. Basal florets disarticulate by complete or partial abscission, the upper florets by fracture; lemmas terminate in two minute teeth; caryopses thick in lower half, the scutellum distinct.

5a. Basal floret disarticulates by complete abscission, the upper florets by fracture.

Lemma dark reddish-brown, coarsely granular, long-pointed; lower florets copiously long-hairy, upper florets sparsely hairy or glabrous; basal hairs long and copious on basal floret, lacking on upper floret, first rachilla segment with a long dense tuft of hairs near the top, lacking on upper floret -------------- *A. sterilis* var. *maxima*, wild red oat (pl. XIV, 36).

5b. Basal floret disarticulates by partial abscission, the upper floret by fracture.

Lemma essentially smooth and glabrous, short-pointed; basal hairs in two dense tufts, not continuous across the back, sometimes lacking; all florets awnless or the basal floret with a geniculate twisted awn, the bend at about the tip of the lemma, sometimes straight; rachilla segment glabrous -------------------------- *Fatuoid*, heterozygous (pl. XII, 32).

4b. All florets of spikelet disarticulate by fracture.

6a. Lemma terminates in two minute teeth, awns if present straight or geniculate and twisted below, the bend usually below the tip of the lemma; rachilla segment glabrous; caryopsis not markedly thickened in lower half, the scutellum large, distinct as in pl. XIV, 37 *A*.[1]

7a. Lemma finely or coarsely granular; rachilla segment usually broken off at or near the base of the basal floret.

A. byzantina, red oat (pl. XI, 28–29).

7b. Lemma essentially smooth; rachilla segment usually entire or nearly so on the basal floret ------------ *A. sativa*, oat (pl. XI, 30).

---

[1] As a result of breeding programs in recent years it is not always possible to classify a variety of cultivated oats definitely as either *byzantina* or *sativa*.

6b. Lemma linear, terminates in two long, stout bristles (not hyaline edged), the nerves light-colored and distinct; all florets awned, the awns geniculate and twisted below; basal hairs few or lacking; rachilla hairs long, confined to the sides on upper half of rachilla---*A. strigosa*, sand oat (pl. XIII, 34).

DANTHONIA, OATGRASS

Spikelets of *Danthonia* are several-flowered and disarticulate above the glumes and between the florets. The seed unit in seed samples is a single floret.

Danthonia species commonly produce cleistogamous spikelets in the lower sheaths of the culms. These florets sometimes occur with crop seeds. They are long and slender, the lemma tip is not cleft, and the rachilla is prolonged into a long bristle (pl. XXVI, 127).

The genus is represented here by one species.

Danthonia spicata, poverty oatgrass (pl. XXVI, 126).—Lemmas are flared out toward the top. A stout, twisted and geniculate awn arises between the acute lobes of the deeply bifid apex. The lemma may be sparsely pubescent or nearly glabrous and ciliate on the margins at the base. Paleas are broad, the apex truncate, hairs on the keels very fine and dense, rachilla and callus are glabrous. Length of lemma 3½–5 mm., width at the top 1½ mm.

DESCHAMPSIA, HAIRGRASS

Spikelets of *Deschampsia* are 2-flowered, disarticulating above the glumes and between the florets. The seed unit is usually a single floret. The genus is represented here by two rather widely distributed species.

Deschampsia caespitosa, tufted hairgrass (pl. XXVI, 128).—Lemmas are silvery, smooth and lustrous, about 2½ mm. long, ½ mm. wide; awn from near the base, straight or semi-geniculate; basal hairs long and spreading at the ends of callus, shorter across the middle; palea bulged up about midway; rachilla about half the length of the lemma, copiously long-pubescent.

Deschampsia flexuosa, crinkled hairgrass (pl. XXVI, 129).—Lemmas are pale brown, lighter toward the tip, sparingly hirsute on upper half, length 4 mm., width 1 mm.; awn from near the base, geniculate and twisted; basal hairs copious, long and spreading; palea not bulged up; rachilla with a tuft of long spreading hairs on either side near the top.

HOLCUS, VELVETGRASS

Spikelets of *Holcus* are 2-flowered, strongly compressed laterally, the glumes about equal and longer than the spikelet, length 4–5 mm.

Basal floret of the spikelet is perfect, the upper floret is staminate. Lemmas are whitish, strongly compressed laterally, smooth and glossy. In lateral view the ventral side is straight, the dorsal side is strongly arched below the middle and slopes above to a pointed apex. Length of floret is about 2–2½ mm., width ½–¾ mm.

Two species are of interest in our agriculture: *H. lanatus*, a crop plant, and *H. mollis*, which is classed as a weed. The seed unit may be either a spikelet or a floret.

Key to *Holcus*

1a. Glumes granular, sparingly short-pubescent and ciliate on the keel nerves; lemma of basal floret awnless, the upper floret with hook-like awn; rachilla of basal floret threadlike, glabrous, lies flat against the palea; rachilla on upper floret lacking; callus essentially glabrous.
Holcus lanatus, velvetgrass (pl. XXVII, 156).

1b. Glumes essentially glabrous, ciliate on the keel nerves; lemma of basal floret awnless, the upper floret with geniculate twisted awn about 3 mm. long; rachilla of basal floret flat and sparsely long-pubescent; the upper floret copiously long-pubescent around the base, often with a pubescent rudimentary rachilla.............. *Holcus mollis*, German velvetgrass (pl. XXVII, 157).

KOELERIA, KOELERIA

Spikelets of *Koeleria* are 2- to 4-flowered, disarticulating above the glumes and between the florets. The glumes are both pointed. Seed unit is usually a floret.

Koeleria cristata, junegrass.—Lemmas are approximately 3 mm. long, ½ mm. wide, granular and opaque, compressed laterally and folded at the tip, short-hispid across the top and pubescent with soft appressed hairs below, the hairs longer along the keel nerve.

Paleas are delicate, hyaline, fully exposed and spread away from the lemma. The rachilla and callus are sparingly pubescent.

The caryopses are yellowish, flattened laterally, thin and semi-soft, 1½ mm. long and ½ mm. wide.

SPHENOPHOLIS, WEDGEGRASS

Spikelets of *Sphenopholis* are 2- to 3-flowered, disarticulating below the glumes. The first glume is short, slender, and pointed. The second glume is longer, broad and obtuse at the top. Articulation is below the glumes and the seed unit may be an entire spikelet.

Sphenopholis obtusata, prairie wedgegrass.—Lemmas are short-pointed, folded at the tip, essentially glabrous and awnless, 2½–3 mm. long, ½ mm. wide or less.

Paleas are delicate, hyaline, fully exposed, spread away from the lemmas as in *Koeleria*. Hairs on the keels of the palea are so short and fine they are scarcely visible at a magnification of 20 ×. The rachilla is sparsely pubescent. The callus is glabrous.

TRISETUM, TRISETUM

Spikelets of *Trisetum* are usually 2-flowered, disarticulating above the glumes and between the florets. The seed unit is usually a floret.

Trisetum flavescens, yellow oatgrass.—Lemmas are 4–5 or 6 mm. long, mostly thin and delicate, compressed laterally, the hyaline, folded apex terminating in two bristles about ½ mm. long. Awns are weak, geniculate and twisted, arising well above the middle of the lemma.

Paleas are delicate, hyaline, fully exposed, tend to spread away from the lemmas, giving the florets a fluffy or chaffy appearance. Callus and rachilla are copiously pubescent with long spreading hairs.

The caryopses are yellowish, flattened laterally, thin and semi-soft, 2½–3 mm. long, ½ mm. wide. (pl. XXX, 203.)

4. Agrostideae—Timothy Tribe

Inflorescence is an open or spike-like panicle. Spikelets are 1-flowered with no sterile or rudimentary florets. Rachilla segments are lacking, rudimentary, or prolonged as a slender bristle in some genera.

Key to Genera of *Agrostideae*

1a. Spikelets disarticulate below the glumes, the glumes equal to or exceeding the floret.
  2a. Glumes compressed laterally, awnless, the midnerve keeled and long-ciliate---------------------------------------*Alopecurus* (p. 37).
  2b. Glumes not compressed laterally or keeled on the midnerve; awns weak, about twice the length of the glumes; glumes minutely hispid and villous on the margins, length 2½ mm---------------*Polypogon* (in part) (p. 39).
1b. Spikelets disarticulate above the glumes.
  3a. Florets compressed laterally-------------------------*Sporobolus* (p. 40).
  3b. Florets not compressed laterally.
    4a. Florets rotund or top-shaped.
      5a. Lemmas and paleas hardened, fitting tightly on the grain; florets obovate, with long-pointed base-----------------*Oryzopsis* (p. 39).
      5b. Lemmas and paleas not as above.
        6a. Pericarp not fused to seed coat, hard, smooth, globose, about 2 mm. in diameter-------------------------*Sporobolus* (in part) (p. 40).
        6b. Pericarp fused to seed coat; caryopsis about 1½ mm. long, pale brownish, roughened by a minute reticulum--------*Phleum* (p. 39).
    4b. Florets not rotund; lemmas rounded or flattened on the back.
      7a. Lemmas awned from near the base, the awns geniculate and twisted.
        8a. Basal hairs copious, spreading, about half the length of the lemma; length of lemma 3–3½ mm----------------*Calamagrostis* (p. 38).
        8b. Basal hairs lacking or only a small tuft at each end of the callus.
          9a. Length of lemma 2 mm. or less----*Agrostis* (in part) (pp. 35–37).
          9b. Length of lemma 3–6 mm., varies with the species. *Alopecurus* (in part) (pp. 37–38).
      7b. Lemma awned from the tip or near it, or awnless; length of lemma 1–2 mm. or 5 mm. in *Aristida*.
        10a. Lemmas terminate in three stout awns-----------*Aristida* (p. 38).
        10b. Lemmas with one awn, usually straight and weak, or awnless.
          11a. Lemmas brownish, dull; rachilla prolonged as a minute bristle. *Apera* (p. 38).
          11b. Lemma whitish, silvery or yellowish.
            12a. Lemmas dull, sparsely short-pubescent--*Muhlenbergia* (p. 39).
            12b. Lemmas glossy or lustrous, glabrous.
              13a. Ventral side bulged out------------*Polypogon* (pp. 39–40).
              13b. Ventral side flattened------*Agrostis* (in part) (pp. 35–37).

AGROSTIS, BENTGRASS

Spikelets of *Agrostis* are 1-flowered, and the rachilla is usually not prolonged. When present it may vary from fairly long and hairy to a mere prickle, as shown in plate VI, 18. These rudimentary rachillas appear to have no diagnostic value. They occur frequently in Highland bentgrass (*A. tenuis* var. Highland) and only rarely in other species.

Eight species and two varieties are described. Four of the species (*A. elliottiana, hiemalis, scabra, exarata*) may occur as incidental weeds with crop seeds but never in quantity.

Seeds of *Agrostis* are too small to be identified accurately without magnification. An examination of a small quantity of seed with a good hand lens or a low-power binocular microscope will show certain features that will give a clue to the identity of the seed bulk, but for the determination of individual seeds that make up the bulk sample a magnification of about 40× is usually necessary.

The descriptions in the seed key that follow apply to normally well-developed seeds and some allowance should be made for natural variations. Size and color may vary somewhat due to stage of maturity or to climatic conditions. Seeds that normally have a glossy or lustrous lemma may be dull under some conditions.

Diagnostic Features of *Agrostis*

Shape.—This can be determined by referring to the illustrations in dorsal view and laying an imaginary horizontal line above the constriction at the base of the lemma. It will be observed that in redtop (*A. alba*) the sides tend to be parallel, producing an oblong-shaped seed of fairly uniform thickness throughout. The lemma is smoothly rounded or slightly flattened on the back. Creeping bentgrass (*A. palustris*) presents a distinct ovate shape. The mass of the seed is concentrated below the middle and the lemma is humped up over the thickest part of the seed. Seeds of Highland bentgrass (*A. tenuis* var.) are usually elliptic in outline, tapering about equally to apex and base. The mass of the seed is concentrated about midway and the lemma is smoothly rounded across the back. Colonial bentgrass (*A. tenuis*) and the variety Astoria are predominantly lance-shaped, the lemma rounded on the back. The bulk of Astoria bentgrass is distinguished from colonial bentgrass chiefly by the larger size of the seeds. However, Astoria bentgrass normally carries a variable proportion of small seeds which would be doubtfully distinguishable from colonial bentgrass.

Awns.—In redtop (*A. alba*) and creeping bentgrass (*A. palustris*), the midnerve may be exserted near the top of the lemma as a short straight awn or only an awn point, but never as a geniculate twisted awn arising near the base of the lemma as in the colonial bentgrass group (*A. tenuis*).

Unlike colonial bentgrass, the variety Astoria produces many seeds with short straight awns arising above the middle of the lemma. Such seeds could be confused with redtop (*A. alba*).

Paleas.—The length of palea and character of the palea tip are other important features. The paleas of some species are relatively long while in other species they are so reduced as to appear to be lacking. The seeds in the illustrations were selected to show these details clearly. As a rule the paleas are not so well exposed and it may be necessary to examine this feature carefully. It is not practical to dissect many seeds for this examination. With experience the identification can be made without exposing the entire palea.

Key to *Agrostis*

1a. Paleas from one-half to slightly exceeding the caryopsis; caryopsis hard.
  2a. Lemmas glossy, pale to deep gold color, sometimes silvery in *palustris;* awns infrequent, if present short and straight, arising well above the middle of the lemma; paleas mostly firm, long and loose.
    3a. Lemmas oblong or lance-shaped, rounded or slightly flattened on the back, the callus smoothly curved; paleas taper evenly from near the base, the apex truncate or with a broad, shallow notch. (pl. VII, 21,*C*).
A. alba, redtop (pl. V, 16).

3b. Lemmas ovate, the back humped up above the constricted base; callus thick and obtusely angled; paleas broad toward the top (shoulder-like), tapering abruptly to a minutely notched or rounded apex (pl. VII, 21, *D*). *A. palustris*, creeping bentgrass (pl. V, 17).

2b. Lemmas grayish, dull or slightly lustrous in Highland bent, occasional lemmas sparingly long-pubescent; awns infrequent, if present geniculate and twisted, arising from near the base of the lemma.

4a. Lemmas elliptic, smooth-textured; basal hairs mostly lacking, if present may be fairly long; paleas mostly long and loose, taper evenly from the base to a sharp-pointed, narrow and deep V-notch (pl. VII, 21, *A*). *A. tenuis* colonial bentgrass var. Highland (Dryland browntop) (pl. VI, 18).

4b. Lemmas lance-shaped, porous textured; basal hairs various or lacking; paleas relatively short, thin, often wrinkled and partially adherent to the grain, the apex variously notched, the points blunt (pl. VII, 21, *B*). *A. tenuis*, colonial bentgrass (browntop) (pl. VI, 19). *A. tenuis* colonial bentgrass var. Astoria (pl. VI, 20).

1b. Paleas reduced to minute transparent scales, usually obscured by the margins of the lemma; caryopses soft or semi-fluid; lemmas dull, whitish, transparent or translucent except in *exarata*.

5a. Nerves of lemma whitish with coarse granules, prominent to the base; awns long, thread-like, arising about ⅔ above the base; basal hairs long, fringe the entire callus............ *A. elliottiana*, Elliott bentgrass (pl. IX, 25).

5b. Nerves of lemma without granules, not prominent to the base.

6a. Lemmas obovate, silvery, opaque, the margins usually overlapping at the top on the ventral side, awned or awnless.

Awnless seeds smooth, length 1½ mm., width ½ mm. Awned seeds coarsely granular and hispid on the nerves at the top; awns long, geniculate and twisted or only awn points, arising about the middle of lemma; length 2 mm., width ½–⅔ mm..... *A. exarata*, spike bentgrass (pl. IX, 26).

6b. Lemmas elliptic or oblong, with fine nerves evident at the top and obscure or lacking below.

7a. Basal hairs at ends of callus long and copious.

Lemma only slightly longer than the grain, so finely granular it appears smooth and semi-polished above the callus; awns infrequent, if present straight and delicate, arising well above the middle of lemma. *A. scabra*, ticklegrass (pl. VIII, 23).

7b. Basal hairs at ends of callus short and stubby or lacking.

8a. Lemma markedly longer than the grain and falling in folds between the four fine nerves at the top, the nerves scabrous; awns long, geniculate and twisted, arising about the middle of lemma, or 5-nerved and awnless in certain strains. *A. canina*, velvet bentgrass (pl. VIII, 22).

8b. Lemma equal to or shorter than the plump grain; awnless; basal hairs rare............ *A. hiemalis*, winter bentgrass (pl. IX, 24).

ALOPECURUS, FOXTAIL

The 1-flowered, strongly compressed spikelets of *Alopecurus* disarticulate below the glumes. The unit commonly considered the seed consists of the ripened floret with its attached glumes. The glumes are equal to the lemma or nearly so, united at the base, and fringed on the keel nerves with long ciliate hairs, except in *myurosoides*.

Mature florets without the attached glumes are sometimes found in processed seed. Lemmas are smooth and translucent and bear a rather weak awn from near the base. The lemma completely infolds the flattened grain so the palea may appear to be missing.

Hulled caryopses are sometimes found with crop seeds. The flattened grains are golden yellow or brownish, soft or semi-fluid, and minutely wrinkled or microscopically striate. The length may vary from 1½ to 3 mm.

Five species are described. *A. pratensis,* meadow foxtail, is in limited cultivation and four other species are field weeds.

Key to *Alopecurus*

1a. Callus at the base large, thick, and prominent; spikelets 5–6 mm. long, 1¾–2 mm. wide; awns of lemma geniculate and twisted.
Glumes pointed, stiff and granular, short-hispid on keel nerves above, long-pubescent on all nerves at the base.
A. myosuroides, slender foxtail (pl. XXIV, 89).

1b. Callus small and inconspicuous; glumes long-ciliate on keel nerves and sparingly long-pubescent on lower half.

2a. Glumes pointed; spikelets 5–6 mm. long, 1¾–2 mm. wide; awns of lemma geniculate and twisted........*A. pratensis,* meadow foxtail (pl. XXIV, 90).

2b. Glumes obtuse.

3a. Spikelets 2 mm. long, 1 mm. wide.

4a. Awns of lemma straight, scarcely exceeding the glumes.
A. aequalis, short-awn foxtail (pl. XXIV, 86).

4b. Awns of lemma geniculate, the bend usually above the glumes.
A. carolinianus, Carolina foxtail (pl. XXIV, 87).

3b. Spikelets 3 mm. long, 1½ mm. wide; awns of lemma geniculate, the bend commonly obscured by the glumes.
A. geniculatus, water foxtail (pl. XXIV, 88).

APERA, APERA

Spikelets of *Apera* are 1-flowered, disarticulating above the glumes, the rachilla prolonged into a very short bristle. The seed unit found with crop seeds is the single floret.

Apera spica-venti windgrass (pl. XXIV, 101).—Lemmas are pale yellowish brown, sparingly short-hispid across the top and terminate in a slender awn, 6–10 mm. long. Lemmas are 2–2½ mm. long, and about ½ mm. wide. The grain is yellowish and semi-soft.

ARISTIDA, THREE-AWN

Spikelets of *Aristida* are 1-flowered, disarticulating above the glumes. The lemma is stiff, completely infolding the grain and usually terminates in three stiff awns, with a sharp-pointed callus at the base. The seed unit in commercial seed is a single floret.

Forty species are reported for the United States. The genus is represented here by one species.

Aristida dichotoma, poverty three-awn (pl. XXIV, 102).—Lemmas are 5–6 mm. long, ¼ mm. wide, dull grayish brown, often flecked with dark brown or black, short-hispid on the midnerve and pubescent across the callus and on the marginal nerves just above the callus. Awns are hispid, the central awn 6–10 mm. long, the two lateral awns shorter.

CALAMAGROSTIS, REEDGRASS

Spikelets are 1-flowered, disarticulating above the glumes. The seed unit in commercial seed is a single floret. Twenty-nine species are recorded for the United States. The genus is represented here by one species.

Calamagrostis montanensis, plains reedgrass.—Lemmas are 3½–4 mm. long, thin and papery, minutely hispid, a geniculate awn from near the base. Rachilla is prolonged into a long-pubescent bristle. The callus is long-pubescent, the bristly pubescence may reach to within ¼ of the top of the lemma.

MUHLENBERGIA, MUHLY

Spikelets of *Muhlenbergia* are 1-flowered, disarticulating above the glumes and the rachilla segment is not prolonged.

Seventy species are recorded for this country. Many of the western species are important range grasses. The genus is represented here by one species. This is an objectionable weed that may occur in lawn grass seed.

Muhlenbergia schreberi, nimblewill.—Lemmas are 3-nerved, minutely hispid and with a few long white hairs at the base. The length is about 2 mm., the width ½ mm. Paleas are bulged out, the keel nerves hispid. Caryopses are dark brown and striate, about 1½ mm. long.

ORYZOPSIS, RICEGRASS

Spikelets of *Oryzopsis* are 1-flowered, disarticulating above the glumes. Lemmas are hardened and fit tightly on the grain, with a slender deciduous awn about 4 mm. long. The seed unit is a single floret. Two species are of agricultural interest.

Oryzopsis miliacea, smilo (pl. XXVIII, 165).—Color is pale yellowish gray. Lemmas are smooth and glabrous, about 2 mm. long and ¾ mm. wide at the widest point. The callus is small and inconspicuous.

Oryzopsis hymenoides, Indian ricegrass (pl. XXVIII, 164).—Color is dark grayish brown to almost black. Lemmas are 3–4 mm. long and 1½ mm. wide at the widest point, with a microscopic reticulum and long white hairs along the margins at the top and on each side of the callus at the base. The hairs may be partly or completely rubbed off in processed seed. The callus, a large shallow cavity with a distinct rim, lies in an oblique position with reference to the long axis of the seed.

PHLEUM, TIMOTHY

The spikelets of *Phleum* are 1-flowered, disarticulating above the keeled glumes. The seed unit is either a single floret or the hulled caryopsis. One species is an important cultivated crop.

Phleum pratense, timothy (pl. XXVIII, 178).—Lemmas are thin and papery, sparsely short-pubescent and 7-nerved above. Paleas are bulged out, the keel nerves close together. Caryopses are about 1½ mm. long, rotund, the embryo projecting at the base of the grain. The surface is distinctly reticulate, the color pale brownish.

POLYPOGON, POLYPOGON

Spikelets of *Polypogon* are 1-flowered, disarticulating below the glumes. The seed unit may be a spikelet or a floret. One species is widely distributed as a weed in cultivated fields.

Polypogon monspeliensis, rabbitfootgrass (pl. XXIX, 185).—Glumes are equal, papery in texture, short-hispid on lower half and villous along the margins and across the callus. Glumes are 2–2½ mm. long and terminate in weak awns 6–8 mm. long. Lemmas are smooth and glossy, about half as long as the glumes, with a delicate awn about 1–1½ mm. long. The paleas are thin and transparent, bulged out in upper half.

The presence of the awn distinguishes seeds of this species from the less widely distributed *Agrostis semiverticillata*. In the absence of the awn, the two appear to be indistinguishable.

SPOROBOLUS, DROPSEED

Spikelets of *Sporobolus* are 1-flowered, compressed laterally or globose in *heterolepis*, and disarticulate above the glumes. Lemmas are 1-nerved, awnless, thin and translucent in many species. Paleas are prominent, equal to or longer than the lemma, often markedly so. In some species the palea splits at maturity, giving the impression of an extra lemma.

The caryopsis falls from the lemma and palea at maturity in many of the species. Thus the seed unit may be either a single floret or a hulled caryopsis. Cleistogamous seeds are commonly produced late in the season in the swollen sheaths of the culms. These caryopses are usually slimmer and longer than those produced from open-pollinated florets.

The genus *Sporobolus* is one of the few grasses in which the pericarp is not firmly fused with the seed coat. In most species it is very thin and closely envelops the grain but slips off quite readily when moist.

Thirty species are reported for the United States. The genus is represented here by eight of the more widely distributed species or species that are in cultivation or in experimental plantings.

The appearance of caryopses may be affected by the stage of maturity or climatic factors. Individual caryopses of closely related species, such as *S. contractus* and *S. cryptandrus*, may not always be distinguishable.

Key to *Sporobolus*

1a. Caryopsis globose, about 2 mm. in diameter, yellowish or pale brown, the pericarp hard and opaque; palea slightly longer than the lemma, splitting at maturity.................................*S. heterolepis*, prairie dropseed.

1b. Caryopses flattened laterally, the pericarp thin and transparent.

2a. Lemma and palea dull, variously mottled with dark brown or black, sparsely pubescent with soft appressed hairs; caryopses light brown, translucent, smooth or microscopically striate, embryo about half the length of the grain.

3a. Floret 3½ mm. long; palea ½–1 mm. longer than the lemma; caryopsis 2–2½ mm. long, ¾–1 mm. wide....*S. vaginiflorus*, poverty dropseed.

3b. Floret 10–12 mm. long; palea about 4 mm. longer than the lemma; caryopsis 2½–3 mm. long, ¾–1 mm. wide.. *S. clandestinus*, scratch dropseed (pl. XXX, 199).

2b. Lemma and palea not mottled or pubescent.

4a. Caryopses light brown, spotted with darker brown and faintly striate; embryo ¾ the length of the grain or longer.

5a. Lemma and palea short-pointed, length 2–2½ mm., width 1–1¼ mm.; caryopsis 1½ mm. long, 1 mm. wide. *S. neglectus*, puff dropseed (pl. XXX, 201).

5b. Lemma and palea long-pointed, length 5 mm., width 1½ mm.; caryopsis 1½–2 mm. long, 1 mm. wide............*S. asper*, tall dropseed.

4b. Caryopses not spotted with darker brown.

6a. Caryopses dark brown, distinctly striate.

Lemma and palea short-pointed, length 2–2½ mm., width 1 mm., the palea splitting at maturity; caryopsis 1½ mm. long, 1 mm. wide, the embryo half the length of the grain or less. *S. airoides*, alkali sacaton.

6b. Caryopses yellowish or yellowish brown, not striate; lemmas and paleas silvery, tissue-like, the paleas commonly splitting at maturity.

7a. Floret short-pointed, 2 mm. long, 1 mm. wide; caryopsis yellowish, not translucent, 1¼ mm. long, ¾ mm. wide.
S. contractus, spike dropseed.

7b. Floret long-pointed, 2 mm. long, ½ mm. wide; caryopsis brownish, translucent, 1 mm. long, ½ mm. wide.
S. cryptandrus, sand dropseed (pl. XXX, 200).

5. Zoysieae—Curly-Mesquite Tribe

Inflorescence is spike-like. Spikelets are mostly 1-flowered in fascicles of 2–5, or single in *Zoysia*. The fascicles fall entire at maturity. Lemmas and paleas are thinner than the glumes. Two genera are of interest in agriculture: *Hilaria*, an important range grass, and *Zoysia*, a lawn grass.

HILARIA, HILARIA

The seed unit in *Hilaria* consists of a cluster or fascicle of three spikelets with their attached glumes. The middle spikelet contains the grain. The two lateral spikelets, each usually 2-flowered, are staminate or sterile. The glumes may be thin and papery or stiff and scabrous. The fertile lemmas are lustrous, thin and delicate in texture. (Refer to p. 8, fig. 5, for structural features.)

Key to *Hilaria*

1a. Fertile or middle floret longer than the glumes; basal hairs long, copious and spreading; glumes thin and papery.

2a. Glumes of lateral spikelets wider at the top, the nerves spreading out fan-shape; caryopses 3-angled, the embryo area extending to about the middle of the grain.............................*H. mutica*, tobosagrass.

2b. Glumes of lateral spikelets acute-pointed, usually with a single awn.
H. jamesii, galleta.

1b. Fertile or middle floret shorter than the glumes; basal hairs short, appressed, or lacking in processed seed.
Glumes of sterile spikelets firm, scabrous, united below, the outer lobe widened toward the top, the inner lobe much narrower; caryopses strongly compressed, the embryo area almost as long as the grain.
H. belangeri, curly mesquite.

ZOYSIA, ZOYSIA

Spikelets of *Zoysia* are 1-flowered, laterally compressed and disarticulate below the glumes. The first glume is lacking, the second glume is stiff, hard and lustrous, short-awned or awn-pointed and completely infolds the thin lemma and palea. The palea is sometimes reduced or lacking. The seed unit is a spikelet.

The species of *Zoysia* are valuable lawn grasses in certain areas and improved strains have been developed. Two species are described here.

Zoysia japonica, Japanese lawngrass (pl. XXX, 204)—Seeds are broadly ovate, short-pointed or short-awned, color dull brownish, commonly with a purplish tinge; length 3½–4 mm., width 1½ mm.

Zoysia matrella, manilagrass (pl. XXX, 205)—Seeds are narrowly elliptic or lance-shaped and short-awned, color bright yellowish straw-color; length 3–3½ mm., width 1 mm.

6. Chlorideae—Grama Tribe

The arrangement of the spikelets in one-sided spikes is one of the outstanding features that is common to the genera included in the grama tribe.

The spikelets are variously modified. For our purpose the genera may be placed into the following two groups with the indicated characteristics: (1) *Chloris* and related genera in which the spikelets consist of one perfect floret with one or more modified or rudimentary florets above it, the spikelets falling entire at maturity; (2) *Eleusine*, *Leptochloa*, and certain other genera which have several-flowered spikelets and resemble some of the genera in the fescue tribe, page 9. Seven genera are included in this study.

Key to the Genera of *Chlorideae*

1a. Spikelets 1-flowered with one or more rudimentary florets above it; lemmas compressed laterally.
  2a. Lemmas awned from the bifid apex; in side view the dorsal side is usually strongly arched ---------- *Chloris* (p. 43).
  2b. Lemmas 3-nerved, the nerves extended into awns; in side view the dorsal side is mostly straight, the ventral side arched ---------- *Bouteloua* (p. 43).
1b. Spikelets without rudimentary florets above the fertile floret.
  3a. Spikelets 1-flowered.
    4a. Spikelets disarticulate below the glumes; the glumes are equal, compressed laterally and inflated ---------- *Beckmannia* (p. 42).
    4b. Spikelets disarticulate above the glumes.
      5a. Lemmas strongly compressed laterally, the rachilla prolonged as a slender bristle; dorsal side strongly arched, ventral side straight. *Cynodon* (p. 44).
      5b. Lemmas not compressed laterally, narrow, terete, long-pointed, the rachilla not prolonged ---------- *Schedonnardus* (p. 46).
  3b. Spikelets several-flowered, rachilla segments present.
    6a. Lemmas compressed laterally, acute-pointed, 3-nerved; caryopses dark brown with transverse or concentric ridges, shape globose or with a deep groove on the ventral side ---------- *Eleusine* (p. 44).
    6b. Lemmas compressed dorso-ventrally, the apex emarginate or mucronate; apex of rachilla expanded into a disk ---------- *Leptochloa* (p. 45).

BECKMANNIA, SLOUGHGRASS

Spikelets of *Beckmannia* are 1-flowered and disarticulate below the glumes, the spikelet falling entire. The seed unit is the entire spikelet or occasionally the hulled caryopsis.

Beckmannia syzigachne, American sloughgrass (pl. XXV, 106.)—Spikelets are flattened laterally, roundish in outline, 3–3½ mm. long and 3 mm. wide. Glumes are keeled, sparingly short-hispid and somewhat inflated, sometimes with faint transverse wrinkles. The apex of the fertile lemma may protrude between the glumes at the top.

Lemma of fertile floret is finely granular, dull and glabrous, keeled on the back and terminates in a short awn. In side view the dorsal side is arched below the middle and tapers to a long-pointed apex.

Palea is glabrous and bulged out by the thick grain. The rachilla is lacking or reduced to a short prickle.

Hulled grains are yellow, semi-soft, keeled, about 1½ mm. long and ½ mm. wide.

BOUTELOUA, GRAMAGRASS

The inflorescence of the gramagrasses is a raceme, each spike of which consists of a few to many sessile spikelets arranged in two rows along one side of the rachis. The spikelets are 1-flowered with one or more rudimentary florets above it. The rudiments are variously modified and are usually characteristic of a species.

Five of the more important native range-grass species are described.

Several horticultural varieties have been developed by selection, a few of which are included in the seed key.

The entire spike falls from the axis at maturity in *B. curtipendula* and the closely related *B. filiformis*. Thus, the unit commonly considered the seed in these two species may be an entire spike.

In another group, *B. gracilis*, *hirsuta*, and *eriopoda*, the spikelets fall free from the glumes, the glumes remaining on the rachis. In this group the seed unit may be a spikelet or a single floret. *B. hirsuta* is further characterized by a prolongation of the rachis of a spikelet as a naked point 5–8 mm. long, and the glumes are tuberculate-hirsute on the midnerve, the tubercles black.

These native species of *Bouteloua* normally produce a poor seed-set. Consequently, the multiple units in commercial seed samples carry a high proportion of empty florets.

Key to *Bouteloua*

1a. Fertile lemma sharply keeled on the back, sparsely long-pubescent along the keel and margins, length about 4 mm.; rudiments extend to about half the length of the fertile lemma, the lobes broadly obtuse and spreading.

2a. Tip of rachilla segment densely long-pubescent; nerves of fertile lemma exserted as short awns; awns of the rudiment equal to those of the fertile lemma or nearly so; caryopsis linear-oval, sharply keeled.
B. gracilis, blue grama (pl. XXV, 108).
(Including the var. Marfa blue grama.)

2b. Tip of rachilla segment not pubescent; fertile lemma deeply 3-cleft, the lobes often winged and terminating in awns about 1 mm. long; awns of rudiment equal or exceed the fertile lemma; caryopsis obovate, not keeled.
B. hirsuta, hairy grama (pl. XXV, 109).

1b. Fertile lemma not keeled on the back, length 5–6 mm.; rudiments equal to fertile lemma or nearly so, the lobes not broad and spreading.

3a. Rudiments reduced to three long bristles; rachilla segment about half the length of the fertile lemma, with a dense tuft of hairs at the tip; callus densely pubescent.
Fertile lemma purplish, inrolled, obscuring the palea.
B. eriopoda, black grama.

3b. Rudiments not reduced to bristles; rachilla segment much reduced, not pubescent at the tip; fertile lemma whitish, not inrolled; callus glabrous or sparsely pubescent on the ends.

4a. Rudiment deeply 3-lobed, the median awn equal to or slightly exceeding the fertile lemma; midnerve of fertile lemma a fine line usually evident to the base............*B. curtipendula*, side-oats grama (pl. XXV, 107).
(Including the vars. Coronado, Vaughn, Uvalde, Encinos.)

4b. Rudiments not deeply lobed, long-awned, the median awn about twice the length of the fertile lemma; fertile lemma short-awned, with a shallow groove on each side of the midnerve..*B. filiformis*, slender grama.

CHLORIS, FINGERGRASS

Spikelets of *Chloris* are several-flowered, and disarticulate above the glumes. The lower floret of the spikelet is sessile and perfect, the upper florets are sterile. Sterile florets consist of empty lemmas, and if more than one is developed the smaller ones are often enclosed in the lower, producing an inflated, wedge-shaped rudiment. The spikelet without the glumes is the unit usually considered the seed.

Two species will serve to represent the genus: *C. gayana*, rhodesgrass, a pasture plant of the Southern States, and *C. virgata*, feather fingergrass, a widely distributed weed found in certain crop seeds.

Fertile lemmas of both species are 3–3½ mm. long and 1 mm. wide. The lemmas are compressed laterally, with a shallow groove between the keel and marginal nerves.

Key to *Chloris*

1a. Keel of fertile lemma strongly humped above the middle; keel and marginal nerves densely villous, the hairs bristly and spreading at the top; awns of rudiment and fertile lemma about equal; color tawny or brownish.
C. virgata, feather fingergrass.

1b. Keel of fertile lemma arched but not strongly humped as above; keel nerves glabrous or sparsely pubescent; awn of rudiment shorter than the awn of fertile floret; color yellowish........*C. gayana*, rhodesgrass (pl. XXV, 121).

CYNODON, CYNODON

Spikelets are 1-flowered, disarticulating above the glumes, the rachilla prolonged as a slender bristle about half the length of the palea. The florets are compressed laterally, the dorsal side strongly arched, the ventral side nearly straight. The seed unit is a spikelet without the glumes or a hulled caryopsis.

Lemmas are firm, the upper half of the keel nerve pubescent.

The genus is represented by the species *Cynodon dactylon*, common bermudagrass, and another form, giant bermudagrass. The taxonomic status of giant bermudagrass is undetermined at the present time and further study is needed. Seed samples of common bermudagrass sometimes contain an admixture of giant bermudagrass. Both are valuable pasture and lawn grasses in some areas but under some conditions they are troublesome weeds. Examination of seed samples suggests that seeds of the two kinds can be separated.

On the basis of our study of the seeds, it appears that the two kinds may be separated on size, color and surface configuration as follows:

Cynodon dactylon, common bermudagrass (pl. XXV, 122).—Florets are about 2 mm. long, ¾–1 mm. wide. The caryopses are light brown and microscopically striate, the apex obtusely pointed. When viewed over a diaphanoscope they appear yellowish and semi-translucent. The length is about 1–1½ mm., the width ½ mm.

Cynodon sp., giant bermudagrass.—Florets are about 2½ mm. long, 1 mm. wide, often purplish. The caryopses are dark grayish brown, dull, with coarse longitudinal striations which are sometimes evident with a 6× hand lens. The apex tends to be acute-pointed. When viewed over a diaphanoscope they appear opaque, or if semi-translucent the color is reddish in contrast to the light yellowish color of common bermudagrass. The length is about 1½ mm., the width 1 mm.

ELEUSINE, ELEUSINE

Spikelets are few to several-flowered, disarticulating above the glumes and between the florets. The seed unit is usually a hulled, dark brown caryopsis, sometimes enclosed in a thin whitish pericarp. It may occasionally be a floret.

The genus is represented here by two species. *Eleusine indica*, goosegrass, is a common field weed. *E. coracana*, African millet, is reported to be used for food in some parts of Africa and Asia. The seeds may occur incidentally with crop seeds.

Key to *Eleusine*

1a. Caryopses compressed laterally, with a deep longitudinal depression on the the ventral side; surface finely striate and transversely ridged; length 1–1½ mm., width 1 mm.

Fertile floret broadest below the middle; lemma thin and glabrous, acute-pointed, the keel nerve minutely winged; palea broad, tends to spread away from the thick grain............*E. indica*, goosegrass (pl. XXVI, 139).

1b. Caryopses globose, the embryo area flattened; surface roughened by minute tubercles and concentric ridges..*E. coracana*, African millet (pl. XXVI, 138).

LEPTOCHLOA, SPRANGLETOP

Spikelets are 2 to several-flowered, disarticulating above the glumes and between the florets. The seed unit is a floret without the glumes, and occasionally a hulled caryopsis. The genus is represented by two species. Both may occur incidentally with crop seeds.

Caryopses for both species are yellowish brown, semi-translucent, the surface faintly striate. The hilum is a prominently elevated, round area with a distinct rim. (See fig. 9.) Embryo is nearly

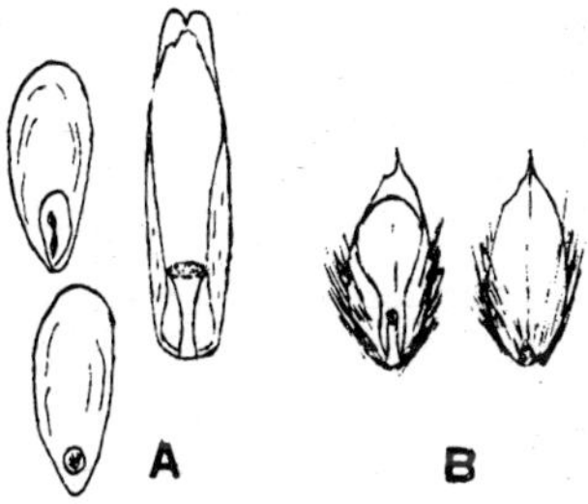

FIGURE 9.—Florets and caryopsis of two species of *Leptochloa*: (A) *L. dubia*; (B) *L. uninervia*.

half as long as the grain, the root-shoot axis a ridge the entire length of the scutellum.

Leptochloa dubia, green sprangletop.—Lemma is glabrous, oblong, notched at the apex and scarcely narrowed at the broad base. The length is about 2½ mm., width 1 mm. The rachilla is slender with the apex expanded into a prominent disk. There is a wide space between the rachilla and the margins of the lemma at the broad base of the floret.

Leptochloa uninervia, Mexican sprangletop.—Lemma is ovate, the apex is broad and mucronate, the base obtusely pointed, marginal nerves are pubescent toward the base. The rachilla is scarcely expanded at the apex. Margins of lemma nearly touch the rachilla at the base.

600705 O—62——4

SCHEDONNARDUS, TUMBLEGRASS

The 1-flowered spikelets disarticulate above the glumes and the rachilla is not prolonged. The seed unit is a spikelet without the glumes. The genus is represented here by one species.

Schedonnardus paniculatus, tumblegrass (pl. XXIX, 186).—Florets are 3–4 mm. long, ½ mm. wide, cylindrical in shape. Lemmas are acuminate and terminate in a short awn, 3-nerved, hispid on the thick midnerve toward the top and sparingly appressed-pubescent below. The base of the floret is broad, the callus in a vertical position. The palea is almost entirely exposed over the thick grain. The keel nerves are close together and exserted at the top as short awns.

7. Phalarideae—Canarygrass Tribe

Inflorescence is a panicle. The 1-flowered, laterally compressed spikelets consist of a pair of enlarged glumes with one perfect terminal floret and two sterile or staminate, often scale-like, florets below it. *Phalaris paradoxa*, with the sterile florets lacking, is an exception. The spikelets disarticulate above the glumes, and the two sterile or staminate florets remain attached to the fertile floret. Lemmas of the fertile florets are glossy or lustrous and more or less pubescent, sometimes glabrous. The seed unit may be a spikelet or a fertile floret.

The tribe is represented here by four genera. For illustration of glumes refer to A. S. Hitchcock's Manual of the Grasses of the United States, U.S.D.A. Misc. Publ. No. 200, 1950.

Key to Genera of *Phalarideae*

1a. Lemmas of sterile lower florets reduced to small scales, much shorter than the fertile floret, or lacking in *paradoxa* .... *Phalaris* (p. 47).

1b. Lemmas of staminate or neuter lower florets equal to or exceeding the fertile floret.

- *2a.* Lemmas of staminate lower florets equal to the fertile floret; golden brown, minutely hispid and villous on the margins, spread apart and exposing the fertile floret .... *Hierochloe* (p. 47).
- *2b.* Lemmas of the sterile lower florets longer than the fertile floret.
  - *3a.* Lemmas of sterile florets equal in length, light brown, awned and pubescent with long, tawny hairs .... *Anthoxanthum* (p. 46).
  - *3b.* Lemmas of sterile florets unequal in length, completely enclosing the fertile floret; pale straw-color, awn-pointed, sparsely pubescent, the hairs long and silky .... *Ehrharta* (p. 47).

ANTHOXANTHUM, VERNALGRASS

Spikelets disarticulate above the unequal glumes and the seed unit is a spikelet with the attached sterile lemmas, or sometimes a fertile floret. (Refer to p. 4 for structural features of the spikelet.)

Two species are described. Both species may occur incidentally with crop seeds. *A. odoratum*, in limited use as a crop, is sometimes included in meadow grass mixtures. Some seeds of the two species may be indistinguishable.

Anthoxanthum odoratum, sweet vernalgrass (pl. XXIV, 99).—Sterile lemmas are about 3 mm. long, coarsely granular, reddish brown with lighter tips, copiously pubescent with long tawny hairs, the margins densely ciliate. The hairs emerge to well above the middle of the lemma. One sterile lemma bears a straight awn from about the middle, the other sterile lemma bears a stout geniculate and twisted awn from near the base. The bend in the geniculate awn usually comes to about the tip of the lemma or only slightly higher. The fertile lemma is about 2 mm. long, awnless, dark reddish brown, glabrous and glossy, and completely encloses the plump grain and palea.

Anthoxanthum aristatum, annual vernalgrass. (pl. XXIV, 100).—Seed of this species differs from sweet vernalgrass chiefly in the position of the pubescence on the sterile lemmas and the relative lengths of the geniculate awns. The points of attachment of the hairs usually do not extend higher than about the middle of the sterile lemmas. The bend in the geniculate awns is usually well above the tip of the lemmas. Both sterile and fertile lemmas tend to be lighter in color than those of sweet vernalgrass.

EHRHARTA, VELDTGRASS

Veldtgrass is one of the more recently introduced range grasses. The seed unit is a spikelet without the glumes.

Ehrharta calycina, veldtgrass.—Spikelets are compressed laterally, 6–7 mm. long. Lemmas of both sterile and fertile florets are pale straw-color and lustrous.

The two sterile lemmas completely enfold the fertile floret. One lemma is narrower and shorter than the other. Both lemmas are sparingly pubescent with long silky hairs and terminate in short awn-points.

Lemma of the fertile floret is flattened laterally, glabrous or with a few silky hairs on the midnerve at the top. The palea is narrow and hyaline and tends to spread away from the lemma. The caryopsis is laterally compressed, minutely roughened, and pale yellowish brown with a dark line on the ventral edge.

HIEROCHLOE, HIEROCHLOE

One species is described. The plant is reported to have some economic uses and the seed occurs occasionally with crop seeds. The seed unit is a spikelet without the glumes, or occasionally only the fertile floret. (Refer to Phalarideae p. 46 for structural features of the spikelet.)

Hierochloe odorata, sweetgrass.—Spikelets are compressed laterally, 4–4½ mm. long. Lemmas of staminate lower florets are equal to the fertile floret, golden brown with lighter hyaline tips, minutely hispid and densely villous along the margins and across the callus. The staminate florets tend to spread apart, exposing the fertile floret. Lemma of fertile floret is golden brown, sparsely hispid across the top and glabrous below.

PHALARIS, CANARYGRASS

Spikelets are laterally compressed and disarticulate above the glumes. A spikelet, except as will be noted, consists of one perfect terminal floret and two sterile lemmas below it. The sterile lemmas in some of the species are inconspicuous and appressed to the fertile lemma like two minute scales. (See fig. 5, p. 8.) Fertile lemmas are stiff and glossy, shorter than the glumes, and enclose the paleas completely.

Nine species are described. Three of these are agricultural crops and six species may occur as field weeds. The seed unit may be an entire spikelet, or it may be a spikelet without the glumes.

Glumes, when present, provide a quick means for identification of the species. However, the glumes may be removed in processing and other features must be considered.

The seed keys are arranged in two parts: (1) Identification by glume character, and (2) identification of spikelets without the glumes.

Key to Glumes of Spikelet of *Phalaris*

1a. Glumes of spikelet broadly winged on the keel.
  2a. Glumes with a prominent tooth-like wing below the sharply pointed apex; length 7–8 mm.......................... *P. paradoxa*, hood canarygrass.
  2b. Glumes without tooth-like wings.
    3a. Wings of glumes taper abruptly to a short point.
      4a. Glumes 7–8 mm. long.................. *P. canariensis*, canarygrass.
      4b. Glumes about 6 mm. long.... *P. brachystachys*, short-spike canarygrass.
    3b. Wings of glumes taper evenly to a pointed apex.
      5a. Glumes 4–6 mm. long................ *P. minor*, little canarygrass.
      5b. Glumes 5–6 mm. long........... *P. caroliniana*, Carolina canarygrass.
1b. Glumes of spikelet not winged on keel or only very slightly so.
  6a. Glumes 5–6 mm. long.
    7a. Entire glumes minutely scabrous...... *P. lemmoni*, lemmon canarygrass.
    7b. Entire glumes not scabrous.
      8a. Keel nerves scabrous, lateral nerves and internerves smooth and glabrous.................. *P. tuberosa* var. *stenoptera*, hardinggrass.
      8b. Keel nerves scabrous, lateral nerves glabrous, the internerves microscopically pubescent............... *P. arundinacea*, reed canarygrass.
  6b. Glumes 3½–4 mm. long; all nerves minutely scabrous, the internerves sparingly pubescent across the top....... *P. angusta*, timothy canarygrass.

Key to Spikelets of *Phalaris* Without the Glumes

1a. Sterile lemmas lacking.
  Fertile lemma elliptic, glabrous, pale gray, often yellowish at top and base; length 3½–4 mm., width 1½–1¾ mm.
  P. paradoxa, hood canarygrass (pl. XXIX, 184).
1b. Sterile lemmas present.
  2a. Sterile lemma one, or if second is present so reduced it appears to be lacking; length about 1 mm.
    3a. Fertile lemma ovate, short-pointed, grayish brown, length about 3 mm., width 1½ mm.......... *P. minor*, little canarygrass (pl. XXVIII, 181).
    3b. Fertile lemma ovate, long-pointed, length 3½–4 mm., width 1½ mm., pale straw-color or brownish.
      4a. Sparsely soft-pubescent across upper portion.
      P. tuberosa var. *stenoptera*, hardinggrass (pl. XXVIII, 182).
      4b. Sparsely soft-pubescent over entire surface.
      P. lemmoni, lemmon canarygrass.
  2b. Sterile lemmas two.
    5a. Sterile lemmas pubescent.
      6a. Sterile lemmas copiously long-pubescent, length about 1 mm.
        Fertile lemma narrowly ovate, grayish brown, sparingly pubescent and ciliate on the margins toward the top; length 3–3½ mm., width 1 mm........... *P. arundinacea*, reed canarygrass (pl. XXVIII, 180).
      6b. Sterile lemmas very sparsely long-pubescent; fertile lemma ovate, sparingly pubescent.
        7a. Fertile lemma tapers from about midway to a long-pointed apex; length 2½–3 mm., width 1–1¼ mm.; sterile lemmas 1–1½ mm. long......................... *P. caroliniana*, Carolina canarygrass.
        7b. Fertile lemma tapers gradually from near the base to a long-pointed apex; length 3½–4 mm., width 1¼ mm.; sterile lemma 1½–2 mm. long...... *P. angusta*, timothy canarygrass (pl. XXVIII, 179).

5b. Sterile lemmas not pubescent; fertile lemmas elliptic, short-pointed, copiously pubescent on upper portion.

8a. Sterile lemmas 3–4 mm. long, loose and spreading, usually broken off in processing.

Fertile lemma pale straw-color, 5–6 mm. long, 2–2½ mm. wide. *P. canariensis*, canarygrass (pl. XXVIII, 183).

8b. Sterile lemmas ½ mm. long or less, not loose and spreading; fertile lemma yellowish brown, 4–5 mm. long, 1½–2 mm. wide. *P. brachystachys*, short-spike canarygrass.

8. Oryzeae—Rice Tribe

Spikelets are 1-flowered, oblong or oval, strongly compressed laterally, sparingly pubescent and more or less rough to the touch. Glumes may be much reduced or wanting. Lemmas and paleas are equal or subequal, both are keeled and terminate in stout awn-points. The lemma of cultivated rice may be short- or long-awned, depending on the variety. The seed unit is usually an entire spikelet. The rice tribe is represented here by three kinds of plants, the cultivated rice (*Oryza sativa*) and the variety red rice, and rice cutgrass (*Leersia oryzoides*). Red rice and rice cutgrass may occur as weeds in rice fields. See fig. 6, p. 8.

Key to Species of *Oryzeae*

1a. Glumes present, much reduced; lemma rigid and brittle, roughened by minute tubercles and scattered appressed hairs; length 7–10 mm., width 3½–4 mm., thickness about 2 mm.

2a. Grain whitish, ---------- *Oryza sativa*, rice (pl. XXVII, 163).

2b. Grain reddish ---------- *Oryza sativa*, red rice.

1b. Glumes lacking; lemma very flat, finely granular, scarcely rough to the touch, keels of lemma and palea bristly-ciliate with scattered short hairs between the keels; length 5 mm., width 1½–2 mm., thickness about ½ mm.; grain reddish brown, finely roughened ---------- *Leersia oryzoides*, rice cutgrass.

9. Paniceae—Millet Tribe

Spikelets of the millet tribe are 1-flowered. Unless otherwise stated, they consist of one perfect terminal floret, a sterile lemma and two glumes below it. The glumes and sterile lemma are thin and tissue-like. The fertile lemma is hardened and may be rough or smooth. It is characterized by a crescent-shaped line just above the callus. It is at this point that the radicle of the germinating seed emerges. (Refer to fig. 6, tribe 9, p. 8, for structural features of the spikelet.) Spikelets disarticulate below the glumes. The seed unit in processed seed is usually an entire spikelet, or in some kinds it may be a hulled caryopsis.

Seeds from 1-flowered spikelets usually do not vary as greatly in size as do those from several-flowered spikelets. However, all measurements stated in the seed keys should be considered approximate only. The stated widths represent the widest point of the seed. Shape refers to the outline of the seed, preferably in dorsal view.

It is not practical to measure the thickness of the seeds. The lateral or edge view of the seed will usually indicate a characteristic shape and the distribution of the mass of the seed; that is, whether the thickness is uniform throughout its entire length or concentrated in certain areas.

The diagnostic features can be recognized under a magnification of 6× or 7×; or 20× may be required for some of the finer details. Eleven of the more important genera are described.

Key to Genera of *Paniceae*

1a. Spikelets flattened dorso-ventrally.
  2a. Spikelets with a prominent knob or cup-like structure at the base. *Eriochloa* (p. 52).
  2b. Spikelets without a knob at the base.
    3a. Nerves on second glume and sterile lemma thick and prominent, with stripes of silky hairs between the nerves; fertile lemma brown, lustrous, finely tubercled. *Leptoloma* (p. 52).
    3b. Nerves and pubescence not as above.
      4a. Spikelets embedded in a flattened, corky rachis; first glume short. *Stenotaphrum* (p. 51)
      4b. Spikelets free from the rachis; first glume short, minute or wanting.
        5a. Second glume and sterile lemma as long as the spikelet. *Axonopus* (p. 50) *Digitaria* (in part) (p. 51)
        5b. Second glume narrow, about half the length of the spikelet. *Digitaria* (in part) (p. 51)
1b. Spikelets not flattened dorso-ventrally, or if flattened the spikelet is thick.
  6a. Caryopsis thick, top-shaped or oblong, protruding from the lemma and palea; lemma and palea villous along the margins; glumes stiff, much reduced. *Pennisetum* (p. 55).
  6b. Caryopsis plano-convex, not protruding from the lemma and palea; glumes and sterile lemma thin and papery.
    7a. First glume much reduced or lacking; second glume and sterile lemma about equal. *Paspalum* (p. 54).
    7b. First glume present, usually much shorter than the second glume.
      8a. Second glume equal to or exceeding the fertile floret. *Echinochloa* (p. 51). *Panicum* (in part) (p. 53). *Setaria* (in part) (p. 56).
      8b. Second glume one-half to two-thirds the length of the fertile floret. *Setaria* (in part) (p. 56). *Panicum* (in part) (p. 53). *Cenchrus* (out of bur) (p. 51).

AXONOPUS, CARPETGRASS
STENOTAPHRUM, ST. AUGUSTINE GRASS

Three species of carpetgrass occur in this country: *A. affinis*, *A. compressus*, and *A. furcatus*. All three are southern species, confined mostly to the coastal regions of the southeast. *A. affinis* is used mainly for pasture purposes. It is the more widely distributed of the two species, and seed in commercial channels at the present time is likely to be of this species. *A. furcatus* may occur incidentally with crop seeds.

The similar-appearing *Stenotaphrum* is used as a lawn grass in southeastern coastal areas and some parts of California. It is a poor seeder and is propagated by stolons. It is included here in order to distinguish it from *Axonopus* in the labeling of seed.

Key to *Axonopus* and *Stenotaphrum*

1a. First glume lacking.
  2a. Spikelets 2–3 mm. long, sparsely long-pubescent along the marginal nerves of the second glume and sterile lemma, the midnerve obscure or lacking.
    3a. Glume and sterile lemma obtuse or short-pointed, equaling or only slightly exceeding the fertile lemma. *A. affinis*, carpetgrass (pl. XXV, 104).
    3b. Glumes and sterile lemma long-pointed, exceeding the fertile lemma. *A. compressus*, broad-leaved carpetgrass (pl. XXV, 105).

2b. Spikelets 4–5 mm. long; glume and sterile lemma long-pointed, exceeding the fertile lemma, glabrous, the midnerve distinct.
A. furcatus, big carpetgrass.

1b. First glume present, short; spikelet 4–5 mm. long; second glume and sterile lemma equal, glabrous and long-pointed, exceed the fertile lemma, the midnerve obscure or lacking__________S. *secundatum*, St. Augustine grass.

CENCHRUS, SANDBUR

Spikelets of *Cenchrus* are 1-flowered and consist of a thin first and second glume and sterile lemma, and a firm fertile lemma and palea. One or more of the spikelets are enclosed within a spiny bur.

C. pauciflorus, field sandbur (pl. XXV, 120).—This is a widely distributed and objectionable field weed. The spines of the bur are sharp and straight, and they are microscopically retrosely barbed. The intact burs are not commonly found in processed seed. The seed unit is usually a hulled caryopsis or a single spikelet.

The first glume of the spikelet is much reduced, the second glume and sterile lemma are nearly equal and shorter than the long-pointed fertile lemma. Caryopses are light brown, broadly ovate, 3–3½ mm. long and 2 mm. wide. The embryo area covers most of the dorsal side. In edge view, the dorsal side is straight and the ventral side is strongly arched.

DIGITARIA, CRABGRASS

Seventeen species of *Digitaria* are recorded for the United States. Most of these are relatively unimportant from an agricultural standpoint. Three of the more important weedy species are included here.

Spikelets of the inflorescence are in twos or threes, arranged in two rows on one side of a 3-angled rachis. Glumes of the spikelet and the sterile lemma are thin and papery. The first glume may be minute or lacking, the second glume may be equal to or shorter than the sterile lemma. Fertile lemmas are firm and stiff, with longitudinal rows of very minute tubercles producing a finely striate effect. The seed unit found in processed seed is usually an entire spikelet.

Key to *Digitaria*

1a. Fertile lemma dark brown or black; first glume lacking or obscure; second glume and sterile lemma strongly ribbed, densely pubescent between the ribs, the hairs commonly terminating in microscopic knobs.

2a. Second glume shorter than the spikelet; spikelet 1½ mm. long, 1 mm. wide.
D. filiformis, slender crabgrass (pl. XXVI, 130).

2b. Second glume and sterile lemma equal to the spikelet; spikelet 2 mm. long, 1 mm. wide________*D. ischaemum*, smooth crabgrass (pl. XXVI, 131).

1b. Fertile lemma pale; first glume minute, second glume narrow, densely ciliate, about half as long as the spikelet; sterile lemma strongly ribbed, appressed-pubescent along the margins; spikelet 2½–3 mm. long, 1 mm. wide.
D. sanguinalis, large crabgrass (pl. XXVI, 132).

ECHINOCHLOA, ECHINOCHLOA

The thin glumes and sterile lemmas of the 1-flowered spikelets are sparingly covered with minute hispid hairs and with long spine-like hairs on the nerves. The hairs sometimes arise from papillae. Awns on the sterile lemmas of *E. crusgalli* are variable. The same panicle may produce extremely long awns (up to 35 mm.) as well as mere awn-points.

Fertile florets are plano-convex, so strongly arched it is difficult to grasp the seed with forceps. Lemma and palea are hard, smooth and glossy. The short point at the tip of the lemma usually withers and breaks away leaving a rather blunt apex.

The seed unit in processed seed may be a spikelet or a hulled caryopsis. Size and shape of the fertile floret are the chief features distinguishing the three species. The widths stated in the seed key refer to the widest point of the seed.

Key to *Echinochloa*

1a. Fertile floret broadest about midway, tapering to an equally pointed apex and base.
- *2a.* Length 2 mm., width 1–1¼ mm.; sterile lemma and second glume of spikelet equal, short-pointed............*E. colonum*, jungle-rice (pl. XXVI, 135).
- *2b.* Length 2¾–3 mm., width 2–2¼ mm.; sterile lemma of spikelet long-awned or only awn-pointed, the second glume shorter, acute-pointed. *E. crusgalli*, barnyardgrass (pl. XXVI, 136).

1b. Fertile floret broadest below the middle, tapering to a long-pointed apex and short-pointed base; length 2¾–3 mm., width 2–2¼ mm.; sterile lemma of spikelet short-awned, the second glume shorter, short-pointed. *E. crusgalli* var. *frumentacea*, Japanese millet (pl. XXVI, 137).

ERIOCHLOA, CUPGRASS

Three species of cupgrass may occur rather commonly as field weeds: *E. contracta*, prairie cupgrass; *E. gracilis*, southwestern cupgrass; and *E. punctata*, Louisiana cupgrass (pl. XXVII, 151). The seed unit found in commercial seed is usually an entire spikelet. Spikelets of the three species are so similar it may not be possible to distinguish them in all cases, especially after processing. Hitchcock's "Manual of the Grasses of the United States," U.S. Dept. of Agric., Misc. Pub. No. 200, should be consulted for the distribution of the species.

The following features are applicable to all three species. The first glume of the spikelet is very short and fused to a cup-like swelling at the base of the spikelet, producing a knob-like structure that is characteristic of this genus. Second glume and sterile lemma are about equal, sparsely long-pubescent, and longer than the fertile floret. Fertile lemmas and paleas are finely roughened by minute tubercles. The fertile lemma bears a short hairy awn which is concealed by the longer glume and sterile lemma. Spikelets, exclusive of awns, are 3½–4½ or 5 mm. long, 1¼–1½ mm. wide

LEPTOLOMA, LEPTOLOMA

One species of *Leptoloma* may be expected to occur with crop seeds, *L. cognatum*, fall witchgrass (pl. XXVII, 162).

First glume of the spikelet is minute or lacking. The second glume and sterile lemma are subequal, prominently 3 to 5 or 7-nerved, with a stripe of soft, appressed hairs between the nerves and along the margins. Glumes tend to spread apart at maturity, exposing portions of the brown fertile lemma.

Fertile lemma and palea are brown and finely roughened by minute tubercles. The lighter colored hyaline margins of the lemma overlap the keels of the palea. Spikelets are 2½–3 mm. long, and 1 mm. wide.

PANICUM, PANICUM

The genus *Panicum* is reported to be the largest of all the grass genera. There are 170 species listed for the United States. Many species are valuable range grasses but comparatively few are in cultivation. Twenty-two species and one variety are included in this study, nine of which are on the market or in experimental plantings. The other species described may occur rather frequently with crop seeds, none of them being objectionable weeds except as they occur in quantity.

Spikelets of *Panicum*.—As shown in figure 6, page 8, the spikelets of *Panicum* consist of two thin and papery glumes, a sterile lemma of like texture, and a fertile floret. The lemmas and paleas of the fertile florets are stiff and hard, or at least firmer than the glumes and sterile lemmas. The fertile lemmas and paleas may be smooth and glossy or roughened by fine tubercles and quite dull.

The relative lengths and character of the first and second glumes and sterile lemmas are shown plainly in the illustrations and these features are not included in the seed key. These structures, when present, are important diagnostic features and should be given careful consideration in making an identification.

The seed unit.—In commercial samples the unit may be an entire spikelet or only the fertile floret. Hulled-out caryopses are present occasionally.

The seed key.—The key describes the fertile florets only. Refer to the illustrations for diagnostic features of the glumes and sterile lemmas. The species are arranged in two sections:

Section 1. Lemmas of fertile florets roughened by minute tubercles. (pp. 53-54).

Section 2. Lemmas of fertile florets smooth and glossy (p. 54).

Key to *Panicum*

SECTION 1

Lemma of fertile floret roughened by minute tubercles.

1a. Florets relatively small, 2–2½ mm. long, ¾–1 or 1¼ mm. wide.
- *2a*. Tubercles on lemma form faint transverse ridges; the nerves of lemma obscure.
  - *3a*. Lemma strongly arched, long-pointed; floret thickest about midway and constricted below ---------------------- *P. distans*, no common name.
  - *3b*. Lemma flattened or only slightly arched.
    - *4a*. Lemma terminates in an awn-point---- *P. reptans*, no common name.
    - *4b*. Lemma not awn-pointed-- *P. maximum*, guineagrass (pl. XXI, 64).
- *2b*. Tubercles on lemma so fine they appear like granules; lemma pale, with distinct fine nerves--------------------- *P. hians*, no common name.

1b. Florets relatively large, 3–4 mm. long, 1½–2 mm. wide.
- *5a*. Lemma and palea both strongly arched in upper portion, constricted at the base; tubercles form very coarse transverse ridges; color brownish or tawny------------ *P. fasciculatum*, browntop panicum (pl. XX, 60).
- *5b*. Lemma flattened or only slightly arched; color yellowish.
  - *6a*. Tubercles form fine transverse ridges.
    - *7a*. Floret broadly oval, obtusely pointed; length 3 mm., width 1¾–2 mm. *P. ramosum*, browntop millet (pl. XXI, 68).
    - *7b*. Floret lance-shaped, the base broad, tapering above the middle to a pointed apex; length 4 mm., width 2 mm. *P. texanum*, Texas millet (pl. XXI, 69).

6b. Tubercles fine, close-spaced, not arranged in a definite pattern; callus pointed, the apex of lemma awn-pointed.
8a. Floret narrowly obovate, length 3 mm., width 1½ mm.
P. purpurascens, paragrass (pl. XXI, 67).
8b. Floret oval, length 3–3½ mm., width 1¾ mm.
P. arizonicum, Arizona panicum (pl. XXI, 63).

Key to *Panicum*

SECTION 2

Lemma of fertile floret smooth and glossy.

1a. Florets olive green, brownish or splashed with brown or gray.
2a. Nerves distinct, fine yellowish lines; florets relatively small, length 1 to 1¾ mm., width ½–1 mm.
3a. Scar small, inconspicuous.
4a. Floret flattened, oblong-- *P. dichotomiflorum*, fall panicum (pl. XX, 59).
4b. Floret slightly arched on the back, thickest about midway.
P. capillare, witchgrass (pl. XX, 57).
P. capillare var. *occidentale*, barbed witchgrass
3b. Scar prominent, wing-like, wide (transversely); floret broadly elliptic.
5a. Scar ½ mm. wide (transversely)-- *P. bergii*, Berg's panicum (pl. XX, 55).
5b. Scar 1 mm. wide (transversely).
P. hillmani, Hillman panicum (pl. XXI, 62).
2b. Nerves obscure or lacking below.
6a. Floret top-shaped, bluish green shading to tawny toward the base; length 2½–3 mm., width 1½–2 mm.
P. brachyanthum, no common name (pl. XX, 56).
6b. Floret neither top-shaped nor bluish green.
7a. Floret diamond-shaped, long-pointed, both lemma and palea bulged out about midway; palea and sometimes the lemma streaked or mottled with dark gray; length 3 mm., width 1½ mm.
P. virgatum, switchgrass (pl. XXI, 70).
7b. Floret elliptic or lance-shaped, short-pointed, the lemma with a prominent bulge or hump about midway, variously streaked with dark brown; length 2¼ mm., width 1 mm.
P. antidotale, blue panicgrass (pl. XX, 54).
1b. Florets not green, brownish, or spotted.
8a. Floret approximately 1½ mm. long, ¾ mm. wide; lemma elliptic, the callus narrow (vertically), light straw-color.
P. anceps, beaked panicum (pl. XX, 53).
8b. Floret 3–4 mm. long, 2–2½ mm. wide.
9a. Floret oval, turgid and highly glossy; both lemma and palea strongly arched above; callus narrow (vertically), the scar prominent; color varies with the variety, light gray, straw, reddish, or almost black.
P. miliaceum, proso or hog millet (pl. XXI, 65).
9b. Floret obovate, strongly arched on the back; lemma with a prominent U-shaped depression above the wide, pointed callus, the scar not prominent; light straw-color-- *P. obtusum*, vine-mesquite (pl. XXI, 66).

PASPALUM, PASPALUM

The first glume of the spikelet is much reduced or lacking in most species of *Paspalum* or, more rarely, both glumes may be lacking. The sterile lemma lies against the palea (flat side) of the fertile floret. Lemmas and paleas of the fertile floret are hard and glossy, except as noted in the seed key, and they may be smooth or faintly roughened.

There are about 50 species of *Paspalum* recorded for the United States. Fifteen species are described here. These include the crop species and those species that are most likely to occur with crop seeds.

The seed unit is usually an entire spikelet. The structural features can usually be recognized under the low magnification of a 6 × or 10 × hand lens. The widths stated in the seed key refer to the widest point of the spikelet.

Key to *Paspalum*

1a. Lemma of fertile florets not evidently roughened by tubercles, strongly arched.

2a. Both glumes of spikelet lacking; fertile lemma granular, whitish and dull, strongly 5- to 7-nerved; length 2 mm., width 1 mm.
P. malacophyllum, ribbed paspalum (pl. XXVIII, 171).

2b. Second glume present; fertile lemma smooth and lustrous, broadly oval with three greenish or brownish prominent nerves; length 2 mm., width 1¾–2 mm. ---------- *P. setaceum*, slender paspalum (pl. XXVIII, 174).
P. supinum, longleaf paspalum.
P. stramineum, sand paspalum.

1b. Lemma of fertile florets finely roughened by longitudinal rows of very minute tubercles, producing a faintly striate effect; lustrous or dull.

3a. Fertile lemma dark reddish brown, strongly arched.

4a. Spikelet broadest about midway and sloping to an equally pointed apex and base; sterile lemma (flat) with four short transverse ridges just inside the marginal nerves; length 2½–2¾ mm., width 1–1¾ mm.
P. plicatulum, brownseed paspalum (pl. XXVIII, 173).

4b. Spikelet obovate, the apex broad, the base obtusely pointed; sterile lemma not transversely ridged; length 2½ mm., width 2 mm.
P. boscianum, bull paspalum (pl. XXVIII, 166).

3b. Fertile lemma not dark reddish brown.

5a. Spikelets flattish, never strongly arched; base of fertile floret broad, the apex short-pointed.

6a. Glume and sterile lemma brownish and densely villous along the margins.

7a. Spikelet 2½ mm. long, 1½ mm. wide; the long dense hairs usually arising from minute dark brown papillae.
P. urvillei, vaseygrass (pl. XXVIII, 175).

7b. Spikelets 3–3½ mm. long, 2–2½ mm. wide; the pubescence less dense than above-- *P. dilatatum*, dallisgrass (pl. XXVIII, 167).

6b. Glume and sterile lemma pale, finely pubescent on upper portion and along the sides, the hairs soft and appressed, the midnerve commonly off-center; a rudimentary first glume sometimes present; apex of glume narrows abruptly to a short point, the midnerve producing a minute wing-like tip; length 2½–3½ mm., width 1½–1¾ mm.
P. distichum, wiregrass (pl. XXVIII, 168).

5b. Spikelets strongly arched; shape oval, ovate, or orbicular.

8a. Glume finely short-pubescent, the hairs arising from minute brownish papillae; spikelet obovate, thickened toward the blunt apex; length 3 mm., width 2 mm. ------------ *P. pubiflorum*, hairyseed paspalum.

8b. Glume glabrous or essentially so.

9a. Glume more or less coarsely wrinkled, the nerves and callus rim obscure; spikelet oval or ovate.

10a. Length of spikelet 3 mm., width 1½–2 mm.
P. notatum var. Pensacola bahiagrass.

10b. Length of spikelet 3½ mm., width 2½–2¾ mm.
P. notatum, Common bahiagrass (pl. XXVIII, 172).
(Including the vars. Argentine and Paraguay bahiagrass.)

9b. Glume not wrinkled, the nerves distinct lines.

11a. Spikelet oval or ovate.

12a. Length 3 mm., width 2½ mm.; midnerve commonly terminates in a minute point; callus inconspicuous.
P. laeve, field paspalum (pl. XXVIII, 170).

12b. Length 4 mm., width 3 mm., the apex obtuse; callus a prominent rim- *P. floridanum*, Florida paspalum. (pl. XXVIII, 169).

11b. Spikelet orbicular, 2¾–3 mm. in diameter.
P. circulare, roundseed paspalum.

PENNISETUM, PENNISETUM

Spikelets of *Pennisetum* are solitary or in fascicles of two or more surrounded by an involucre of bristles. The bristles remain attached to the spikelets at maturity. Three species are in cultivation: *P. ciliare*, buffelgrass; *P. glaucum*, pearl millet or cattail millet; and *P. purpureum*, napiergrass.

The fascicles of *P. ciliare* consist of 1–5 spikelets with a persisting involucre of bristles. Outer bristles are short, slender and scabrous; the inner ones are thicker and ciliate and about twice the length of the spikelet. The seed unit in commerce consists of an entire fascicle.

P. glaucum and *P. purpureum* are more widely grown. The first glume of the spikelet is short, sometimes minute or lacking, the second glume is shorter than the sterile lemma. Fertile lemma and palea are hard, stiff and villous along the margins. The thick caryopsis protrudes from the lemma and palea at maturity. Commercial seed consists almost entirely of hulled caryopses, many of which are sometimes seriously damaged in processing. The caryopses have a characteristic color, bluish above and shading to a tawny color toward the base.

There appears to be no constant feature whereby individual seeds of *glaucum* and *purpureum* can be distinguished with certainty. The following observations apply to most of the seeds of each kind and may be helpful in identifying bulk samples.

P. glaucum, pearl millet or cattail millet, including the varieties Starr and Gahi (pl. XXVIII, 176).—Caryopses are obovate and thickened toward the top. In edge view, the dorsal side is straight, the ventral side is arched. Length is 3–3½ mm., width 2–2½ mm.

P. purpureum, napiergrass (pl. XXVIII, 177).—Caryopses are oblong, flattened dorso-ventrally. In edge view both dorsal and ventral sides are essentially straight. Length is 4 mm., width 2 mm.

SETARIA: MILLET; BRISTLEGRASS; FOXTAIL

As in *Panicum*, spikelets of *Setaria* are 1-flowered and consist of two thin and papery glumes, a sterile lemma of like texture, and one fertile floret without a prolonged rachilla. The first glume is usually much reduced. (Refer to fig. 6, p. 8, for these structural features.)

Lemmas and paleas of the fertile florets are hard and stiff and may be glossy or dull, smooth or roughened by fine tubercles or ridges. The sterile lemma usually bears in its axil a reduced, tissue-like palea which often adheres to the fertile palea against which it lies. This is often seen on seeds of foxtail millet, *S. italica*.

The seed unit in commercial seed may be an entire spikelet but usually it is the fertile floret without the glumes and sterile lemma, or it may be a hulled caryopsis. The shape of the fertile floret and the character of the lemma may be considered the most constant diagnostic features.

With the exception of *S. magna*, the portion of the palea between the keel nerves is roughened by minute tubercles. The portion between the keel nerves and the margins is smooth and glossy. In certain species the margins of the lemma extend to the keels of the palea so that the glossy portions are not exposed. In other species the mature grain spreads the lemma and palea apart about midway, leaving crescent-shaped segments of the glossy portions of the palea exposed. However, if such a seed is not fully matured, the grain will not be filled out sufficiently to spread the lemma and palea apart to expose the glossy segments. This possiblity should be kept in mind and all other features taken into consideration when making an identification, especially in such species as *S. faberi*, giant foxtail.

Seven species and several varieties of cultivated millet, *S. italica*, are described. Six of the species are field weeds, one of which, *S. faberi*, is classed as a noxious weed in some States.

Similar-Appearing Species

Seeds of *Setaria viridis* and *S. verticillata* are too similar to be distinguishable. *S. viridis* appears to be the more widely distributed as a field weed, and seeds found incidentally with crop seeds are likely to be of this species.

Plants of *Setaria geniculata* are reported to be quite variable. The seeds may vary in size from those stated in the seed key, and all structural features should be carefully evaluated in order to distinguish the larger seeds from *S. faberi*, giant foxtail.

Setaria italica, Foxtail Millet

Seeds of the varieties of foxtail millet listed in the seed key under *7a* have the same morphological features, differing only in color. Single panicles of the variety Siberian and the variety Hungarian may produce variable proportions of golden yellow seeds. Yellow seeds of these two varieties and the variety Common are doubtfully distinguishable. Such seeds should not be removed from samples of Siberian or Hungarian millet as being another variety. A growing test should be made for accurate determination.

Growing tests indicate that there are two forms of the so-called German foxtail millet. The plants exhibit a difference in panicle type as well as in seed character. One form (seed key *4a*) produces seeds that are uniformly tubercled over the entire lemma. The other form (*4b*) produces seeds that are tubercled over the entire lemma and a proportion of seeds that are tubercled only along the margins, sometimes only faintly so. Such seeds might be confused with the variety Common millet (*7a*).

Key to *Setaria*

1a. Lemma roughened by minute tubercles over entire surface or only along the margins in *4b*.

2a. Tubercles not arranged in a definite pattern; glossy segments of the palea exposed.

3a. Lemma dull; in edge view thickness about uniform throughout, flattened at the tip; light brown mottled with darker brown or yellowish; length 2 mm., width 1–1¼ mm.; second glume equal to fertile lemma or nearly so, often persisting ---------- *S. viridis*, green foxtail (pl. XXIX, 187).
S. verticillata, bur bristlegrass

3b. Lemma oleaginous; in edge view turgid toward the top, especially in the variety White Wonder, the apex thick and blunt; color golden yellow; length 2–2¼ mm., width 1½ mm.; second glume about ¾ the length of the fertile lemma.

4a. Entire surface of lemma tubercled.
S. italica, German foxtail millet (pl. XXIX, 190).
Including the var. White Wonder (pl. XXIX 197).

4b. Entire surface of lemma tubercled or only along the margins.
S. italica, German foxtail millet (pl. XXIX, 196).

2b. Tubercles arranged in a pattern of transverse ridges; lemma dull, yellowish, greenish or brown, often black-tipped in *geniculata*; second glume ½–⅔ the length of the fertile lemma.

5a. Ridges of lemma wide-spaced, the spaces flat and lustrous, the ridges evident to the tip; glossy segments of palea not evident.

6a. Seeds in edge view thickest about midway and sloping abruptly to a short-pointed apex and base; length 3 mm., width 2 mm.
S. lutescens, yellow foxtail (pl. XXIX, 190).

6b. Seeds in edge view thickest in lower half, tapering above the middle to a long-pointed apex, length 2½ mm., width 1¼ mm.
S. geniculata, knotroot bristlegrass (pl. XXIX, 188).

5b. Ridges of lemma close-spaced, the spaces usually not distinct, the ridges much reduced toward the top, the tip often smooth and lustrous; glossy segments of palea evident in fully developed seeds; in edge view thickest in lower half, tapering above the middle to a long-pointed apex; length 2½ mm., width 1¾ mm.----*S. faberi*, giant foxtail (pl. XXIX, 189).

1b. Lemma not roughened by minute tubercles; glossy segments of palea evident.

7a. Lemma oleaginous, with faint transverse undulations; in edge view thickness of seed fairly uniform throughout, not turgid toward the top; length 2½–2¾ mm., width 1½ mm.; second glume about ¾ the length of the fertile lemma.

Pale golden yellow, *S. italica* var. Common foxtail millet (pl. XXIX, 194). (Includes the var. Goldmine and certain seeds of the vars. Siberian and Hungarian.)

Orange or golden yellow, *S. italica* var. Siberian foxtail millet.

Uniformly orange color, *S. italica* var. Kursk foxtail millet.

Dark brown, light brown mottled with dark brown, or golden yellow.
S. italica var. Hungarian foxtail millet (pl. XXIX, 195).

7b. Both lemma and palea smooth and highly glossy, lemma terminates in a stout awn-point; seed broad and flat, length 2 mm., width 1½ mm.; second glume equal to fertile lemma.
S. magna, giant bristlegrass (pl. XXIX, 192).

10. Andropogoneae—Sorghum Tribe

Structural features.—Spikelets are borne in pairs at each node of the rachis. One spikelet of the pair is sessile and perfect, the other is pedicellate and usually sterile. The sterile spikelet is suppressed in some species, only the pedicel being developed.

Some species may produce a variable proportion of fertile pedicellate spikelets, as in *Andropogon gerardi*, for example. Such spikelets differ in appearance from the sessile spikelets. They tend to be cylindrical in shape and lack the rachis segment and pedicel of other florets.

The inflorescence breaks apart at or near the nodes of the rachis (fig. 6, *C*, p. 8). Each joint or unit consists of a sessile spikelet (*a*), with a segment of the rachis (*b*) and the pedicel of a second spikelet (*c*). The pedicellate spikelet may or may not persist.

Disarticulation of the rachis.—The rachis may break apart by abscission or by fracture. In disarticulation by abscission the base of the sessile spikelet develops a callus and separates smoothly from the rachis on the suture line. The top of the attached rachis segment and of the pedicel of the sterile floret is usually slightly expanded, with a smooth rim around a cup-like depression. This is referred to as a "cupped" apex in the descriptions that follow.

In some species the nodes are lightly solidified and the rachis breaks apart at or near the nodes, the ends appearing rough or splintered as in sudangrass.

The seed unit.—The sessile fertile spikelet with its attached rachis segment and pedicel is the unit commonly considered the seed in commercial samples. (See fig. 6, *C*.) Glumes of the spikelet are usually stiff and hardened and enclose two florets, one of which consists of a tissue-like sterile lemma and the other contains the caryopsis within a tissue-like lemma and palea. Awns, borne on the thin fertile lemmas, are usually rather weak and easily detached.

Commercial seed sometimes carries caryopses that have been hulled from the glumes in processing. The characteristic shapes and sizes of the caryopses are shown in the illustrations of some of the species.

Key to Genera of *Andropogoneae*

1a. Rachis segment and pedicel united throughout; fertile spikelet globose and deeply pitted ---------- *Hackelochloa* (p. 60).
1b. Rachis segment and pedicel not united.
  2a. Second glume of fertile spikelet sharply keeled throughout; first glume flat or with a shallow longitudinal groove.
    3a. Pedicellate spikelet suppressed, only the hairy pedicel developed.
      Sorghastrum (p. 61).
      Andropogon (in part) (p. 59).
    3b. Pedicellate spikelet present, sterile or sometimes fertile in some species of *Andropogon* ---------- *Andropogon* (p. 59).
      Hyparrhenia (p. 60).
  2b. Second glume of fertile spikelet not sharply keeled throughout.
    4a. First glume pointed; spikelets thick and robust -------- *Sorghum* (p. 61).
    4b. First glume broadly winged at the summit; spikelets flattened dorso-ventrally; grain reddish brown, finely striate -------- *Eremochloa* (p. 60).

ANDROPOGON, BLUESTEM

Ten species are included in this study. One species, *A. virginicus*, may occur as a weed contaminant with crop seeds and nine species are important range, hay, or pasture grasses. Improved varieties of these grasses are introduced from time to time, some of which are included in the seed key.

The seed key describes the fertile, sessile spikelets. When fertile pedicellate spikelets are produced, they differ from the sessile spikelets mainly in the terete shape of the grain and in the absence of a rachis segment and pedicel.

Key to *Andropogon*

1a. First glume with a broad longitudinal groove on the back; rachis segment and pedicel flattened below and cupped at the expanded apex; glumes dull brownish, long-pointed, length 7–9 mm., width 1½ mm.
  A. gerardi, big bluestem (pl. XXIV, 91).
  A. hallii, sand bluestem (pl. XXIV, 92).
1b. First glume not grooved on the back; rachis segment and pedicel both cupped at the apex except in *scoparius* and *virginicus* (*8a* and *8b*); glumes pale or with purplish cast except in *scoparius*.
  2a. Both rachis segment and pedicel with a prominent longitudinal groove, the groove purplish or brownish, ⅔–¾ the length of the pointed glumes; spikelet 3–3½ mm. long, 1 mm. wide.
    3a. First glume with a minute pit on the back about midway or higher.
      A. intermedius. Australian bluestem (pl. XXIV, 95).
    3b. First glume not pitted on the back ---- *A. caucasicus*, Caucasian bluestem.
  2b. Both rachis segment and pedicel not grooved.
    4a. Pedicel with a reddish longitudinal groove, rachis segment not grooved, about ¾ the length of the pointed glumes; spikelet 4 mm. long, 1 mm. wide.
      A. ischaemum, yellow bluestem (including the var. King Ranch bluestem) (pl. XXIV, 94).
    4b. Neither pedicel nor rachis segment grooved, about ½ the length of the glumes.
      5a. Glumes flared out toward the top.
        6a. First glume winged and markedly wider than the second glume in upper half, stiffly ciliate on the margins, may be faintly papillose; the wings not reaching the summit form a broad and slightly 3-lobed tip; spikelet 4–4½ mm. long, 1½ mm. wide.
          A. nodosus, angletongrass (including the var. Medio bluestem).

6b. First glume not winged, scarcely wider than the second glume in upper portion.

7a. First glume coarsely papillose-hairy, especially toward the top and along the sides, the hairs long and stiff; glumes often purplish; spikelet 4 mm. long, 1½ mm. wide.
A. annulatus, Diaz or Kleberg bluestem (pl. XXIV, 96).

7b. First glume with a fringe of long stiff hairs (5–6 mm.) across the back near the tip, not papillose; spikelet 3–3½ mm. long, 1 mm. wide............*A. sericeus*, silky bluestem (pl. XXIV, 97).

5b. Glumes long-pointed, rachis segment cupped, pedicel with sterile spikelet much reduced or lacking.

8a. Pedicel thread-like, flexuous, markedly longer than the glumes; rachis segment slender, not expanded at the top; spikelet 3–3½ mm. long, ½ mm. wide......*A. virginicus*, broomsedge (pl. XXIV, 98).

8b. Pedicel and rachis segment stout, shorter than the glumes, flattened below and expanded toward the top; glumes yellowish or brownish; spikelet 6–8 mm. long, 1 mm. wide
A. scoparius, little bluestem (pl. XXIV, 93).

EREMOCHLOA, CENTIPEDEGRASS

Eremochloa ophiuroides, centipedegrass (pl. XXVII, 150).—Centipedegrass is a lawn grass in some areas of the Southern States. As in other genera of the sorghum tribe, the inflorescence breaks up into joints. The length of the joint is 3–3½ mm., width 1½ mm. The glume of the fertile sessile spikelet is smooth and glabrous, broadly winged at the summit and with 5 or 6 spine-like projections on the margins toward the base. The entire joint is rarely present in commercial seed and the seed unit is usually a hulled caryopsis.

The caryopsis is oval, flattened dorso-ventrally, reddish brown and finely striate. The length is about 2 mm., the width 1–1½ mm.

HACKELOCHLOA, LIZARD-TAIL

Hackelochloa granularis, lizard-tail (pl. XXVII, 155).—The inflorescence breaks up into joints, each joint bearing a sessile brownish perfect spikelet and a pedicellate sterile spikelet that is laterally compressed and winged. The sterile spikelet is usually missing in processed seed.

Lizard-tail may occur as a contaminant in certain southern crop seeds. The seed unit is a joint of the rachis. The rachis segment and pedicel are united. The perfect spikelet is globose and deeply pitted, 1½–2 mm. long.

HYPARRHENIA, THATCHGRASS

The seed unit is a joint of the rachis of the inflorescence, 3–4 or 5 mm. long, ¾–1 mm. wide. The first glume of the fertile, sessile floret is flat or slightly sunken and more or less copiously long-pubescent, with stiff and spreading hairs. Rachis segment and pedicel are about ¾ the length of the spikelet, both are flat (not grooved), with long, dense hairs on the edges. The oval rim at the apex of the rachis segment usually has a prominent point at the back. Two species are of experimental interest in the South.

H. rufa, thatchgrass.—Pubescence on the first glume is reddish brown, becoming tawny toward the base. The second glume is sharply keeled and glabrous.

H. hirta, jaraguagrass.—Pubescence on the first glume is whitish and copious. The second glume is not keeled or if keeled only slightly so, and sparsely whitish-pubescent toward the top.

SORGHASTRUM, SORGHASTRUM

Sorghastrum nutans, Indiangrass.—This is a widely distributed native species. The seed unit is a joint of the rachis of the inflorescence, 6–7 mm. long, 1½ mm. wide. The glumes are golden yellow or brownish yellow. The first glume is flat (not grooved), glabrous or with a few long grayish hairs. The second glume is sharply keeled and glabrous.

Rachis segment and pedicel are about half as long as the spikelet, very slender, often purplish, densely pubescent with long spreading whitish hairs, the hairs completely encircling the callus at the base of the floret. Sterile florets are lacking; only the pedicels are developed.

The hulled caryopses sometimes occur with crop seeds. They are reddish brown and finely striate, about 3 mm. long and 1 mm. wide at the top, which is the widest point. The shape is obovate, tapering to a long-pointed base. Compare this with the caryopsis of *Sorghum halepense,* johnsongrass, page 64.

SORGHUM, SORGHUM

Refer to page 58 for structural features of the inflorescence and spikelets of the sorghums.

The grain and forage sorghums are a large group which includes several botanical varieties and a vast number of agronomic varieties. The term "sorgrass" is recognized as the name of a kind. It includes rhizomatous derivatives of crosses of johnsongrass x sorghum or johnsongrass x sudangrass. The identification of the seeds of this group must await future developments. At present *Sorghum almum* is considered a kind and is not classed with the sorgrass group.

It would not be possible to treat the identification of seeds of such a large and complex group adequately within the scope of this paper. Attention will be directed to the identification of seeds of three species which are a problem common to most parts of the country: *S. halepense,* johnsongrass; *S. almum,* sorghum almum; and *S. sudanense,* sudangrass.

The seed unit in commerce for these three species is a joint of the inflorescence (the sessile floret with attached rachis segment and pedicel). Some seed lots carry a proportion of hulled caryopses. Inasmuch as johnsongrass and sorghum almum are classed as noxious weeds in some parts of the country, it is important that they be distinguished from sudangrass both in the spikelet and caryopsis forms.

The descriptions that follow indicate the possibilities for accurate determination of the seeds. It should be possible to distinguish spikelets and hulled caryopses of *Sorghum sudanense* both in bulk and as individual seeds. Spikelets and hulled caryopses of *S. halepense* and *S. almum* are distinguishable when viewed in bulk but the individual seed units are not always distinguishable.

600705 O—62——5

SORGHUM HALEPENSE, S. ALUM, S. SUDANENSE

Spikelets

Glume color and structural features of the spikelets of the three species may be very similar in many cases. The size, shape, and manner of articulation must be evaluated carefully and the possibilities and limitations of making an accurate identification by visual means should be kept in mind at all times.

Glume color of *Sorghum halepense* and *S. almum*.—The color is essentially the same in both species. It is predominantly a glossy mahogany, with a variable proportion of glumes that are light reddish, shading to straw-color above or entirely straw-color.

Glume color of *Sorghum sudanense*.—There are a number of varieties of this species in cultivation and the glume color may vary with the variety. Some varieties are of a uniform color. Other varieties may be a mixture of colors to the extent that they are indistinguishable. This can be illustrated by the following varieties: California-23, uniformly straw-color; Wheeler, straw-color with a small proportion of dark mahogany glumes or reddish shading to straw-color above; Tift, straw-color with a variable proportion mahogany; Piper and Greenleaf, mahogany with a variable proportion of straw-color or reddish, shading to straw.

It is evident, therefore, that glume color may not be a reliable criterion for distinguishing the three species, or the varieties of sudangrass from each other.

Size of spikelets.—As in many kinds of seeds, size is useful mainly in the extreme ranges, that is, a seed is normally very small or very large. The tabulation that follows shows that spikelets of *Sorghum halepense* average the smallest in all three dimensions and those of *S. sudanense* are the largest.

Species	Spikelets		
	Length (mm.)	Width (mm.)	Thickness (mm.)
S. halepense	6 or less	2	1¼–1½
S. almum	6	2¼–2½	1¾
S. sudanense	6–6½	2½–2¾	2

The same proportionate size differences are illustrated in figure 10.

The size difference is apparent when seeds are viewed in bulk and this provides a reliable clue to the identity of a sample. Individual seeds of *S. halepense* and *S. almum* are not always distinguishable by size because of the natural variation within a species. Individual seeds of *S. sudanense* are distinguishable from the other two species by the consistently larger size when considered in connection with the characteristic shape of the spikelet.

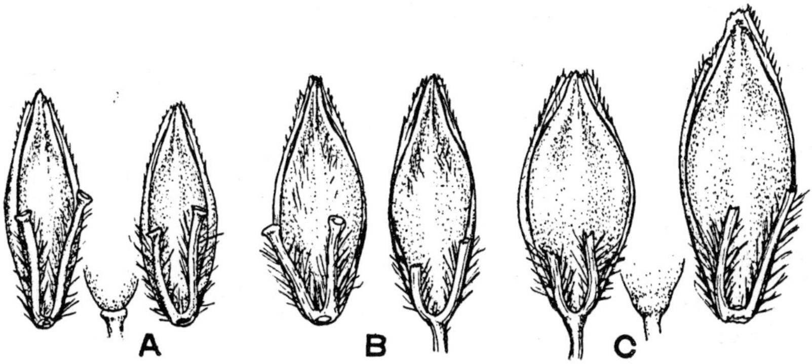

FIGURE 10.—Comparative size and shape of spikelets of three species of *Sorghum* in ventral view: (*A*), *S. halepense;* (*B*), *S. almum;* (*C*), *S. sudanense.* Approximately 5. ×

Shape of spikelets.—The grain in *Sorghum halepense* is usually markedly shorter than the glumes and the spikelet tends to be long-pointed. The grain of *S. almum* is only slightly shorter than the glumes so that the spikelet appears short-pointed and heavier toward the top than in *halepense*. This difference in shape is not very apparent because the tips of the glumes are often damaged in processing.

Spikelets of *Sorghum sudanense* are broadest about midway and taper equally to the apex and base. The grain extends to the tip of the blunt apex of the glumes. The distinctive shape and large size of the spikelets distinguishes this species from *halepense* and *almum*.

Disarticulation of spikelets.—Refer to page 58 for this structural feature, and to figure 10.

There are three points of articulation, the base of the sessile fertile spikelet, the tip of the rachis segment, and pedicel of the sterile spikelet. The manner of disarticulation is by abscission or by fracture as follows.

Sorghum halepense: All parts disarticulate by abscission.

Sorghum almum: Some seeds disarticulate by abscission as in *halepense*, and some seeds by fracture as in *sudanense*.

Sorghum sudanense: All parts disarticulate by fracture.

SORGHUM HALEPENSE, S. ALMUM, S. SUDANENSE

Caryopses

Color of caryopses.—The color of caryopses may vary somewhat but well-developed grains may show a characteristic color as follows: *Sorghum halepense*, dark reddish brown with a lighter brown scutellar area, the entire surface finely striate; *S. almum*, light reddish brown, the scutellar area usually yellowish brown, the surface striate as in *halepense; S. sudanense*, color as in *almum*, the surface not striate.

Size of caryopses.—The tabulation that follows shows a gradation in size similar to the spikelets. The grains of *Sorghum halepense* average the smallest in all three dimensions, those of *sudanense* the largest.

Species	Caryopses		
	Length (mm.)	Width (mm.)	Thickness (mm.)
S. halepense	2–3	1¼–1¾	1
S. almum	3¼–3½	2	1½
S. sudanense	4–4½	2¼–2½	1½–2

The comparative size differences are illustrated in figure 11.

Shape of caryopses (fig. 11).—*Sorghum halepense* and *S. almum* are essentially the same in shape. The caryopses may be obovate with a pointed base, or broadly oval with an obtuse apex and base. In lateral or edge view the thickness of the grain is quite uniform throughout its entire length, the ends thick and blunt.

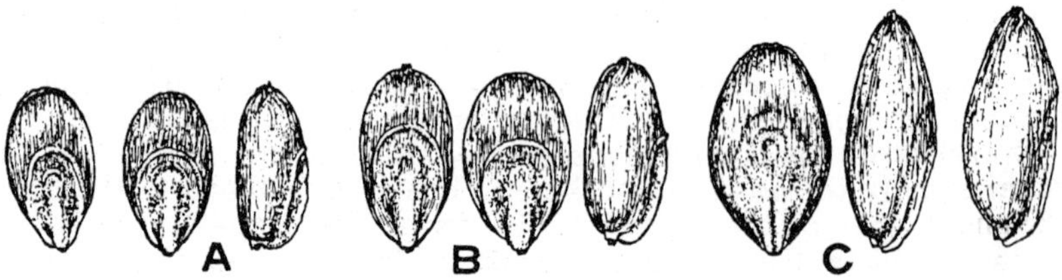

FIGURE 11.—Caryopses of three species of *Sorghum*, showing comparative size and shape in dorsal and lateral views: (*A*), *S. halepense;* (*B*), *S. almum;* (*C*), *S. sudanense.*

The caryopses of *Sorghum sudanense* are broadly elliptic, tapering about equally to a blunt apex and base. In lateral or edge view the grain is thickest through the middle, with both the ventral and dorsal sides curved and the tip of the grain flattened dorso-ventrally as if pinched together.

The embryo (fig. 11).—The embryos of *Sorghum halepense* and *S. almum* are similar. The scutellar area may be a broad oval or it may be slender and pointed toward the base, its margins well-defined. The root-shoot axis appears as a thick ridge along its entire length. A fine line in the form of a loop is sometimes present above the plumule tip.

The margins of the scutellar area in *Sorghum sudanense* are usually poorly defined. The root-shoot axis is flattened at the top and terminates in two raised ridges at the base. A fine raised line in the form of a loop is usually present above the plumule tip.

It is possible to identify caryopses of *Sorghum sudanense* by size, shape and character of the embryo. Caryopses of *S. halepense* and *S. almum* appear identical in all structural features except size. Because of natural variation in seed size, some caryopses of these two species may be indistinguishable.

THE LEGUME FAMILY

Classification of legumes.—Legumes may be divided into three subfamilies: *Mimosoideae, Caesalpinioideae,* and *Papilionoideae.* Some authors treat these as separate families: *Mimosaceae,* mimosa family; *Caesalpiniaceae,* senna family; and *Fabaceae,* pea or bean family. The latter classification is followed here.

Inasmuch as there are comparatively few species in the mimosa and senna families that are of interest either as field crops or weeds, the seeds of the two families are treated here in one group. The pea or bean family contains large numbers of genera and species of important field crops as well as field weeds. The descriptions and illustrations of the seeds in this large group are arranged alphabetically for convenient reference.

The seed unit.—Fruits in these three families may be 1-seeded or several-seeded pods. The several-seeded pods may dehisce or split open at maturity along both sutures and the seed unit in such species is a true seed. This may be illustrated by the pods and seeds of vetch and peas. Indehiscent pods of some species are of such texture that they shell off readily in processing and the seed unit is also a true seed, as in the clovers. Indehiscent pods of other species do not break readily and the seed unit may be an entire fruit as in lespedeza, or a segment of a pod as in serradella, *Ornithopus sativus.*

STRUCTURAL FEATURES OF LEGUME SEEDS

Mimosaceae, mimosa family, and Caesalpiniaceae, senna family.—Unlike the irregular corollas of the pea family, flowers of these two families have regular corollas, or only slightly irregular in some species. The seeds also differ markedly from those of the pea family. They tend to be elongate and flattened, the two faces being plane or only slightly rounded. (See pl. XXXVII, 296.) The hilum is very small, inconspicuous and featureless, and located at one end of the seed. Seeds in cross-section reveal rather thin cotyledons and an abundant endosperm (fig. 12, *A*).

Fabaceae, pea or bean family.—Flowers in this family have irregular corollas, typified by the familiar pea-like flower. Seeds of the different genera and species vary greatly in size, shape and color, and in the location of the hilum and chalaza. Figure 12, *B* and *C*, shows the structural features that are characteristic of most species in this family.

The hilums have a feature that is characteristic of the family. Regardless of size or shape of the hilum, there is a fine longitudinal groove or slit down the middle. (See fig. 12, *C*.) There are a few exceptions in which the slit is obscured by a persisting layer of whitish corky tissue, as in cowpeas (*Vigna sinensis*) and beans (*Phaseolus*).

The chalaza is evident as a small dark-colored area on the surface of many species of legume seeds, near the upper end of the cotyledons. As will be pointed out, this is an important diagnostic feature in a few species, such as vetch.

The micropyle is a minute pore near one end of the hilum, leading down to the nucellus. It appears to have no diagnostic value.

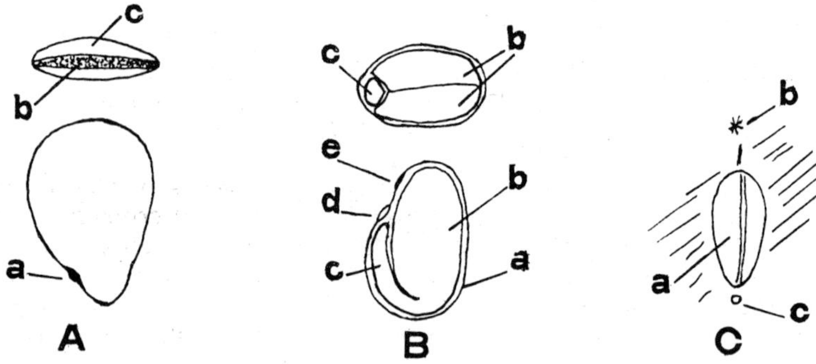

FIGURE 12.—Diagrams of seeds in lateral view and cross-section, showing structural features representative of three legume families. *A*, Mimosa and Senna families: (a) Position of hilum, (b) cotyledons, (c) endosperm. *B*, The pea family: (a) seed coat, (b) cotyledons, (c) radicle, (d) hilum, (e) chalaza. *C*, Detail of hilum area in pea family: (a) Hilum, (b) chalaza, (c) micropyle.

The raphe is evident in certain species as a raised ridge between the hilum and the chalaza. It has no diagnostic significance in the species under consideration.

Seed coats are usually thick and hard and often impermeable to water. There are some exceptions, as peanut (*Arachis*) for example. Here the pod (pericarp) furnishes the protective cover and the brown seed coat is thin and papery.

Unlike seeds of the mimosa and senna families (fig. 12, *A*), the embryos in the pea family fill the entire cavity of the seed and there is no visible endosperm (fig. 12, *B*). The position of the radicle is usually indicated on the seed surface by a slight depression between it and the cotyledons. The hilum lies near the tip of the radicle.

The shape of the seed is determined to a great extent by the angle at which the radicle lies against the cotyledons and the distance between the tip of the radicle and the tip of the cotyledons. The diagrams in figure 13 illustrate some of the forms.

Seed size may be useful only in the extreme ranges, that is, an unknown seed may fall within the limits of a very small-seeded group or a large-seeded group. Because of natural variation, all sizes stated in the keys and descriptions should be considered as approximate only. Unless otherwise specified, the length is determined by a line perpendicular to the hilum. Thus in some cases a seed may be wider than

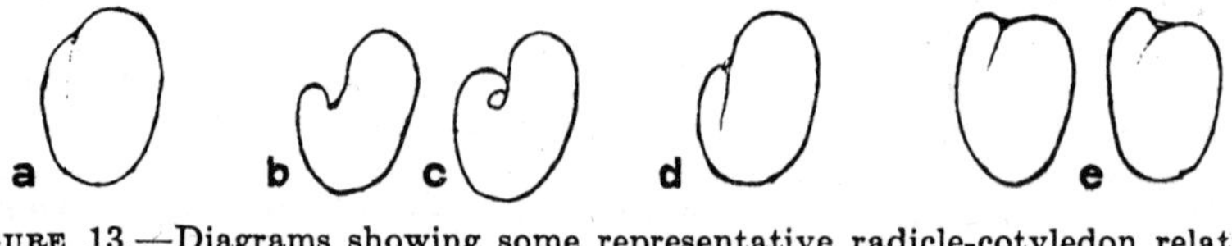

FIGURE 13.—Diagrams showing some representative radicle-cotyledon relationships: (a) Radicle lies close to the edge of the cotyledons throughout; (b) radicle tip strongly divergent from the cotyledons; (c) radicle tip divergent and curved; (d) radicle about half the length of the cotyledons; (e) radicle exceeding the cotyledons.

long. Width represents the widest point of the seed at right angles to the perpendicular dimension.

DISTINGUISHING THE MIMOSA, SENNA, AND PEA FAMILIES

The following gross features may serve as a guide for determining the family of a legume seed.

1. Hilum inconspicuous, minute and featureless, lies at one end of the long axis of the seed; seeds elongate, flat, and usually thin, the two faces plane or only slightly rounded; endosperm present------Mimosa and senna families (p. 65).
2. Hilum may be minute or very large but with a distinct median groove or slit, or in some species the hilum is obscured by a prominent layer of whitish tissue; seed size and shape various, not thin and flat as above; endosperm lacking. Pea or bean family (pp. 65-66).

THE SEED KEYS

Seeds of the legumes are arranged in three keys. The mimosa (Mimosaceae) and senna (Caesalpiniaceae) families are combined in one key (pp. 67-69). This key includes 12 representative genera and species. Seeds of the more important species are shown diagrammatically in figure 14.

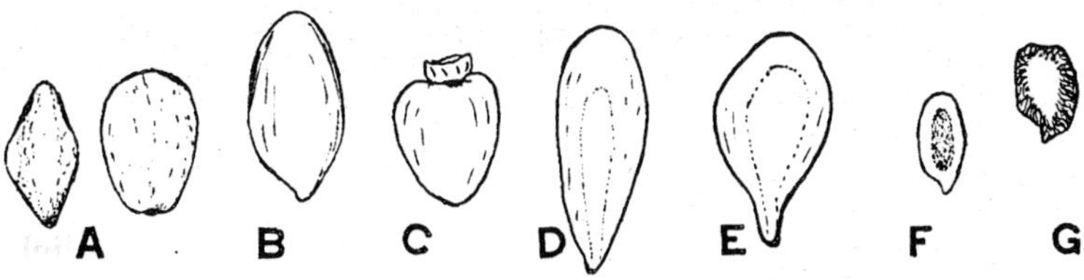

FIGURE 14.—Characteristic features of some representative species in the mimosa and senna families of legumes: *A*, *Ceratonia siliqua* in lateral and dorsal views; *B*, *Gleditsia triacanthos; C*, *Ulex europaeus; D*, *Desmanthus leptolobus; E*, *Desmanthus illinoensis; F*, *Cassia marilandica; G*, *Cassia hebecarpa.*

The second key (p. 69) classifies 10 genera of the pea family (Fabaceae) that have indehiscent pods. These may occur in commercial seed as entire pods or segments of pods. Representative species of these 10 genera are shown diagrammatically in figure 15.

The third key (pp. 69-71) classifies the hulled seeds of 40 genera in the pea family (Fabaceae). These genera have dehiscent pods or, in some species, the pods are indehiscent but shell out readily in processing. Some genera of the latter class may also be listed in the previous key.

KEY TO SPECIES IN MIMOSA AND SENNA FAMILIES

1a. Seed smooth, without surface configurations.
  2a. Length 10–12 mm., width 7mm.
    3a. Color mahogany, highly glossy; broadly oval and usually bulged up at the middle point on each face.
      Ceratonia siliqua, carob or St. Johnsbread (fig. 14, *A*).
    3b. Color chocolate brown, dull or lustrous; oblong or oval, both faces flattened-----------------*Gleditsia triacanthos*, honeylocust (fig. 14, *B*).
  2b. Length less than 10 mm.
    4a. Seeds with a prominent U-shaped caruncle; length exclusive of caruncle 3-3½ mm., width at widest point 2-2½ mm.

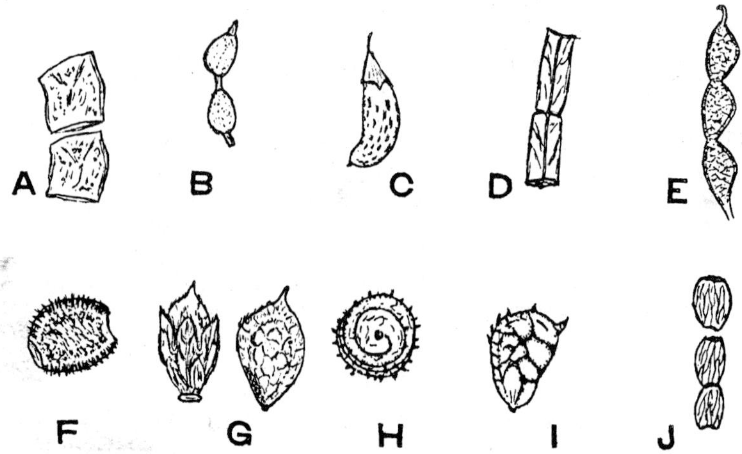

FIGURE 15.—Indehiscent pods of 10 genera of the pea family: *A*, *Aeschynomene; B*, *Alhagi; C*, *Amorpha; D*, *Coronilla; E*, *Desmodium; F*, *Hedysarum; G*, *Lespedeza; H*, *Medicago; I*, *Onobrychis; J*, *Ornithopus.*

5a. Color light to dark reddish brown, glossy; caruncle appears wing-like in side view; shape roughly triangular (broadest at the hilar end and pointed toward the opposite end) or occasionally broadly oval; thickness not uniform throughout, may be humped up on cotyledon portion or otherwise misshapen; radicle usually shorter than the cotyledons. *Ulex europaeus*, gorse (fig. 14, *C*).

5b. Color yellow, yellowish brown or greenish, lustrous; caruncle appears knob-like in side view; shape broadly oblong or oval, the thickness uniform throughout, the sides symmetrically rounded; radicle about equal to the cotyledons.............. *Cytisus scoparius*, Scotch broom.

4b. Seeds without caruncle.

6a. The two faces slightly convex; color light to dark chocolate brown, lustrous; broadly oval, sometimes slightly wrinkled; length 6 mm., width 3½–4 mm........................ *Cercis canadensis, redbud.*

6b. The two faces flattened.

7a. Color greenish brown, dull or lustrous; obovate; length 3½–4 mm., width 2½ mm. *Hoffmannseggia densiflora*, Indian rushpea (pl. XXXVII, 296).

7b. Color grayish brown and faintly spotted with black; roundish, glossy; length and width about equal, 5½–6 mm. *Hoffmannseggia jamesii*, rushpea.

1b. Seed surface variously configured.

8a. Seeds minutely pitted, dark brown; oblong or obovate; length 3½–4 mm., width 2½ mm.......... *Cassia nictitans*, sensitive pea (pl. XXXVI, 281).

8b. Seeds not pitted.

9a. Seeds rectangular, about as thick as wide; chocolate brown and glossy, each face with a yellowish, linear stripe lying diagonally across the seed; length 5–6 mm., width 2½–3 mm. *Cassia tora*, sicklepod (pl. XXXVI, 282).

9b. Seeds not as thick as wide.

10a. A fine U-shaped line on each face of seed.

11a. Broadly obovate, light brown; length 4 mm., width 2½ mm. *Desmanthus illinoensis*, Illinois mimosa (fig. 14, *E*, p. 67).

11b. Linear-obovate or oblong, reddish brown; the narrow U-shaped area within the lines slightly arched; length 5 mm., width 2 mm. *Desmanthus leptolobus*, prairie mimosa (fig. 14, *D*, p. 67).

10b. A prominent obovate or oval area down the middle on each face of the seed; seed dull.

12a. Seeds square or nearly so; a tan-colored striate membrane usually persisting outside of the oval central area; length and width about equal, 5–6 mm., joints of pod about as long as wide.
Cassia hebecarpa, wild senna (fig. 14, *G*, p. 67).

12b. Seeds oblong, 5–6 mm. long, 3½ mm. wide; the tan-colored membrane usually not persisting; joints of pod much wider than long.
Cassia marilandica, wild senna (fig. 14, *F*, p. 67).

KEY TO GENERA IN PEA FAMILY

Genera With Indehiscent Pods

1a. Pods 1-seeded.

2a. Pods with a very coarse reticulum on both faces and a few spine-like projections toward the top on the curved edge.
Onobrychis, sainfoin (fig. 15, I, p. 68; pl. XXXVIII, 324).

2b. Pods not coarsely reticulate.

3a. Pods narrowly oblong, the tip curved upwardly; glabrous or with appressed grayish pubescence and prominent dark brown, resinous elevations--------------------*Amorpha*, false indigo (fig. 15, *C*, p. 68).

3b. Pods obovate, with a distinct but fine reticulum and a short whitish pubescence, the calyx persisting.
Lespedeza, lespedeza (fig. 15, *G*, p. 68; pl. XXXVII, 303).

1b. Pods few- to many-seeded.

4a. Pods coiled, smooth or spiny.
Medicago, alfalfa, bur-clover, buttonclover (fig. 15, *H*; pl. XXXVIII, 315–320).

4b. Pods not coiled, disarticulate on the partitions between the seeds.

5a. Segments of pod terete, surface with obscure longitudinal wrinkles.
Coronilla, crownvetch (fig. 15, *D*, p. 68).

5b. Segments of pod not terete.

6a. Segments rectangular or square.

7a. Segments flattened, each face with anastamosing raised lines.
Ornithopus, serradella (fig. 15, *J*; pl. XXXVIII, 326).

7b. Margins of segments flattened, the central portion elevated and warty-----*Aeschynomene*, joint vetch (fig. 15, *A*; pl. XXXVI, 271)

6b. Segments not rectangular or square.

8a. Segments oval or triangular, flattened.

9a. Faces covered with short spine-like projections.
Hedysarum, sulla (fig. 15, *F*, p. 68; pl. XXXVII, 295).

9b. Faces with a distinct reticulum of fine brownish lines and sparingly pubescent with very short whitish hairs.
Desmodium, beggarweed (fig. 15, *E*, p. 68; pl. XXXVII, 292).

8b. Segments oval and bead-like, both faces rounded and faintly roughened--*Alhagi*, camelthorn (fig. 15, *B*, p. 68; pl. XXXVI, 272).

Genera With Hulled Seeds

Refer to pages 65-67 for structural features of the seeds.

The key is arranged in two sections on the basis of hilum size. Section 1 (p. 69): The hilums are large and prominent and require little magnification. Section 2 (pp. 70–71): The hilums are minute (¼–½ mm. long) and may require a magnification of 10 × or 20 × to show the character for a genus.

Section 1, Hilums Large and Prominent

1a. Hilum broadly oval.

2a. Seed surface smooth.

3a. Width of seed greater than the length; hilum at the base of the upturned end--------------------------*Aeschynomene* (p. 71; pl. XXXVI, 271).

3b. Width and length of seed approximately equal.

4a. Seed variously streaked or mottled with black or brown.
5a. Seeds dull............... *Pisum* (in part) (p. 84; pl. XXXIX, 329). *Lathyrus* (in part) (p. 76).
5b. Seeds highly glossy, marbled with black. *Lathyrus* (in part) (p. 76; pl. XXXVII, 298).
4b. Seeds not streaked or mottled.
6a. Hilum depressed below surface of seed. *Daubentonia* (p. 74; pl. XXXVII, 291).
6b. Hilum flush with seed surface. *Vicia* (in part) (p. 90; pl. XLI, 377). *Pisum* (in part) (p. 84; pl. XXXIX, 329).
2b. Seed surface variously roughened. *Lathyrus* (in part) (p. 76; pl. XXXVII, 299).
1b. Hilum linear, linear-oval or wedge-shaped, depressed or flush with the seed surface.
7a. Width of seed greater than the length.
8a. Hilum obscured by a whitish tissue. *Phaseolus* (p. 83; pl. XXXVIII, 328). *Strophostyles* (p. 85; pl. XXXIX, 333). *Vigna* (p. 92; pl. XLI, 379).
8b. Hilum not obscured.
9a. Hilum with a collar-like structure around the margin. *Stizolobium* (p. 85; pl. XXXIX 332).
9b. Hilum without appendages.... *Glycine* (p. 74; pl. XXXVII, 294). *Lathyrus* (p. 76; pl. XXXVII, 301).
7b. Width of seed not greater than the length.
10a. Hilum approximately 2–3 mm. long.. *Glycine* (p. 74; pl. XXXVII, 294). *Lathyrus* (in part) (p. 76; pl. XXXVII, 300). *Vicia* (in part) (p. 90; pl. XLI, 374).
10b. Hilum linear, about ½–⅔ the length of the circumference of the seed. *Vicia* (in part) (p. 90; pl. XLI, 373).

SECTION 2, HILUMS MINUTE, ¼–½ MM. LONG

1a. Hilum near the middle of the long axis of the seed, scarcely indented.
2a. Seeds mottled, sometimes faintly so. *Alysicarpus* (p. 72; pl. XXXVI, 273). *Lotus* (in part) (p. 79; pl. XXXVIII, 310). *Sesbania* (p. 84; pl. XXXIX, 331). *Pueraria* (p. 84; pl. XXXIX, 330).
2b. Seeds not mottled.
3a. Width 2–4 times the length.
4a. Seeds oblong.................. *Coronilla* (p. 73; pl. XXXVI, 285). *Galega* (p. 74; pl. XXXVII, 293). *Sesbania* (p. 84; pl. XXXIX, 331).
4b. Seeds kidney-shaped, often variously warped. *Medicago* (in part) (p. 81; pl. XXXVIII, 319).
3b. Width only slightly more than the length; seeds ovate. *Desmodium* (p. 74; pl. XXXVII, 292).
1b. Hilum not near the middle of the long axis of the seed.
5a. Radicle equal to the cotyledons or nearly so, or exceeding them.
6a. Seeds mottled................. *Adesmia* (p. 71; pl. XXXVI, 270). *Melilotus* (in part) (p. 81; XXXVIII, 322). *Lupinus* (in part) (p. 80; pl. XXXVIII, 313). *Trifolium* (in part) (p. 85; pl. XL, 354).
6b. Seeds not mottled.
7a. Seed coat smooth.... *Lupinus* (in part) (p. 80; pl. XXXVIII, 314). *Medicago* (in part) (p. 81; pl. XXXVIII, 319). *Melilotus* (in part) (p. 81; pl. XXXVIII, 321). *Lespedeza* (in part) (p. 77; pl. XXXVII, 305). *Trifolium* (in part) (p. 85; pl. XXXIX, 348).
7b. Seed coat not smooth.
8a. Surface pitted................ *Ononis* (p. 83; pl. XXXVIII, 325). *Indigofera* (p. 76; pl. XXXVII, 297).

8b. Surface with blister-like protuberances or warty.
Cyamopsis (p. 74; pl. XXXVII, 290).
Medicago (in part) (p. 81; pl. XXXVIII, 318).
Trifolium (in part) (p.85; pl. XXXIX, 339).
Melilotus (in part) (p. 81; pl. XXXVIII, 323).
Trigonella, in part (p. 90).

5b. Radicle about half the length of the cotyledons, or if longer never equal to them.

9a. Seeds mottled, or only faintly spotted in *Astragalus*.

10a. Radicle tip widely divergent from the cotyledons at the top.
Alhagi (p. 71; pl. XXXVI, 272).
Astragalus (in part) (p. 72; pl. XXXVI, 279).
Crotalaria (p. 73; pl. XXXVI, 288).

10b. Radicle tip not widely divergent from the cotyledons at the top.
---------------------*Lespedeza* (in part) (p. 77; pl. XXXVII, 306).

9b. Seeds not mottled or spotted.

11a. Surface deeply pitted or ridged----------*Trigonella* (in part) (p. 90).
Astragalus (in part) (p. 72; pl. XXXVI, 278).
Indigofera (p. 76; pl. XXXVII, 297).

11b. Surface smooth or nearly so.

12a. Radicle tip widely divergent from the cotyledons toward the top.
Astragalus (in part) (p. 72; pl. XXXVI, 276).
Hedysarum (p. 76; pl. XXXVII, 295).
Crotalaria (p. 73; pl. XXXVI, 286).
Medicago (in part) (p. 81; pl. XXXVIII, 317).

12b. Radicle tip not divergent from the cotyledons.
Lespedeza (p. 77; pl. XXXVII, 303).
Ornithopus (p. 83; pl. XXXVIII, 326).
Anthyllis (p. 72; pl. XXXVI, 274).
Trigonella (in part) (p. 90; pl. XL, 364).
Trifolium (in part) (p. 85; pl. XL, 361).

Descriptions and Keys to Species in the Pea Family

ADESMIA

Adesmia muricata, adesmia (pl. XXXVI, 270).—Seeds are broadly oval, the length and width nearly equal, 1½–2 mm. Color is gray or brownish, copiously mottled with black and slightly lustrous. The radicle is equal to or shorter than the cotyledons and divergent from them at the tip. The hilum is minute and lies in a shallow indentation at the broader end of the seed. The presence of this seed may indicate Argentine origin.

AESCHYNOMENE

Aeschynomene virginica, Northern jointvetch (pl. XXXVI, 271).—Seeds are dark brown, markedly wider than long, length about 2½ mm., width 4 mm. The thick radicle is about ⅓ the length of the cotyledons and widely divergent at the tip, producing the effect of an oblong seed with an upturned end. The hilum is relatively large, about 1 mm. long, concave, and with a distinct light brown rim around it. This species may occur incidentally with such crops as rice.

ALHAGI

Alhagi pseudalhagi, camelthorn (pl. XXXVI, 272).—Seeds are greenish or reddish brown, obscurely flecked with black, 2 mm. long and 3–3½ mm. wide. Shape is variable but tends to be oval. The radicle is about one-half the length of the cotyledons and divergent toward the tip. The hilum is minute and usually lies off-center on the long axis of the seed.

Seeds of camelthorn have been found in alfalfa imported from Turkestan. The plant is locally established in southern California but, insofar as known, it has not been found in domestic seed.

ALYSICARPUS

Alysicarpus vaginalis, alyceclover (pl. XXXVI, 273).—Seeds are oblong; in edge view the seed is broader at the ends than at the middle. Seeds are lustrous yellowish brown and finely stippled with dark purple, occasional seeds greenish or reddish without the stippling. Length is about 1¼ mm., width 2 mm. The hilum is minute, lies at about the middle of the long axis of the seed. The radicle is about half the length of the cotyledons and not divergent from them.

This species is in limited use as a crop in the South.

ANTHYLLIS

Anthyllis vulneraria, kidneyvetch (pl. XXXVI, 274).—Seeds are oval, 2½–3 mm. long, 1½ mm. wide. The color is a distinctive feature, about one-half of the seed is bright green and the other half is yellowish or pale brownish. The radicle is about one-half the length of the cotyledons, the tip not divergent. The hilum is minute, usually with a whitish rim. A prominent brown spot marks the position of the chalaza.

This seed may occur incidentally in imported seeds such as crimson clover or alfalfa.

ASTRAGALUS

The genus *Astragalus*, milkvetch, comprises a great many species. Among them are the poisonous locoweeds, and some species which appear to have forage value. Six representative species are included here.

The species are all characterized by a minute hilum which in most cases lies in a deep indentation at the base of a divergent radicle tip. It will be observed in the illustrations that there is considerable variation in shape of seeds in some of the species. It is difficult to establish a representative size or shape for such species. A size difference between species may be evident when several seeds are viewed but individual seeds may be indistinguishable.

The term "overall length" as used in the seed key represents a perpendicular line from the tip of the cotyledons, instead of from the hilum as is usually done.

Key to *Astragalus*

1a. Seeds symmetrical in shape, the hilum in a relatively broad indentation.
- *2a.* Yellowish brown, broadly oval, the overall length and width nearly equal, 2–2½ mm.; radicle slightly shorter than the cotyledons and widely divergent at the tip........*A. cicer*, chickpea milkvetch (pl. XXXVI, 276).
- *2b.* Dull olive green, kidney-shaped, the overall length 2½–3 mm., width 2 mm., radicle less than half the length of the cotyledons and widely divergent at the tip.......*A. chinensis*, Chinese milkvetch (pl. XXXVI, 275).

1b. Seeds squarish or oblong, not symmetrical in shape, variously bent and misshapen; hilum usually lies in a deep and narrow indentation.
- *3a.* Seeds pitted; color light tan finely stippled with black, yellowish around the hilum..............*A. flexuosus*, flexuous milkvetch (pl. XXXVI, 278).

3b. Seeds not pitted.
4a. Color yellowish brown minutely stippled with black.
A. nuttallianus, nuttall milkvetch (pl. XXXVI, 279).
4b. Color yellowish brown, not stippled with black.
5a. Radicle ½–¾ the length of the cotyledons; overall length 2½–3 mm., width 1¾ mm.----*A. falcatus*, sicklepod milkvetch (pl. XXXVI, 277).
5b. Radicle nearly equal to cotyledons in most seeds; overall length 2 mm., width 1½ mm. (Compare with *Melilotus*, pl. XXXVIII, 321.)
A. rubyi, ruby milkvetch (pl. XXXVI, 280).

CICER

Cicer arietinum, chickpea or garbanzo (pl. XXXVI, 283).—Seeds are roundish in outline, lobed at one end and short-pointed at the opposite end, one side slightly flattened. Length, width, and thickness are about equal, 8–10 mm. The hilum is relatively large and concave, and lies on the flattened side near the pointed end with a large and prominent chalazal spot above it. The surface may be finely roughened by minute blister-like protuberances or ridges. The usual color is whitish but the color may vary with the variety.

Chickpea is in limited cultivation in California.

CORONILLA

Seeds of *Coronilla*, crownvetch, are narrowly oblong, much wider than long, the minute hilum lying in a shallow indentation at about the middle of the long axis of the seed. Color is a dull reddish brown. The length is about 1–1½ mm., width 4–5 mm. Two species are of interest.

Coronilla varia, crownvetch (pl. XXXVI, 285).—This species is a cultivated crop. The seeds may also occur incidentally with other crop seeds.

Coronilla scorpioides, scorpion crownvetch (pl. XXXVI, 284).—Seeds of this species may occur incidentally with imported seeds such as alfalfa or red clover. It may be distinguished from *C. varia* by the fine median line on each side of the seed and the somewhat recurved shape.

CROTALARIA

Four species of *Crotalaria* are utilized to some extent for forage and soil improvement in the Southern States: *C. intermedia*, slenderleaf crotalaria; *C. mucronata* (*C. striata*), striped crotalaria; *C. lanceolata*, lance crotalaria; and *C. spectabilis*, showy crotalaria. The species *spectabilis* and *mucronata* have become troublesome because of poisonous qualities of the seed if eaten by livestock and poultry. Another species, *C. juncea*, sunn crotalaria, does not seed well under most conditions in the South and is not in general use because of insufficient seed.

Key to *Crotalaria*

1a. Radicle over half the length of the cotyledons, divergent, the tip recurved and hook-like, obscuring the minute hilum.
2a. Seeds black and highly glossy, the hilum surrounded by a band of whitish waxy tissue; overall length 4–4½ mm., width 3½ mm.
C. spectabilis, showy crotalaria (pl. XXXVI, 289).
2b. Seeds dark olive green or brownish, dull or lustrous, overall length 6–7 mm., width 4–4½ mm.--------------------------*C. juncea*, sunn crotalaria.

1b. Radicle about half the length of the cotyledons, divergent but the tip not hooked, the minute hilum well exposed.

3a. Seeds light-colored.

4a. Color gold or pale reddish, highly glossy; overall length 2 mm., width 1¾ mm. .... *C. lanceolata*, lance crotalaria (pl. XXXVI, 287).

4b. Color pale tan, lustrous or glossy; overall length 2½–3 mm., width 2–2½ mm. .... *C. intermedia*, slenderleaf crotalaria (pl. XXXVI, 286).

3b. Seeds light gray with irregular concentric streaks of dark gray or dark green; overall length 3½ mm., width 3 mm.
C. mucronata (*C. striata*), striped crotalaria (pl. XXXVI, 288).

CYAMOPSIS

Cyamopsis tetragonolobus, guar (pl. XXXVII, 290).—Seeds of guar are somewhat flattened and circular in outline, 4–5 mm. in diameter. The color is pale gray or yellowish, occasional seeds may be brownish or purplish. Surface is roughened by minute blister-like protuberances. The radicle is equal to the cotyledons and not divergent, the hilum is minute.

The crop is being grown to some extent in the Southwest.

DAUBENTONIA

Daubentonia texana, rattlebox (pl. XXXVII, 291).—The seed is broadly oval and rotund, light reddish brown in color, 5–5¼ to 6 mm. long and 6–7 mm. wide. Radicle is equal to the cotyledons and not divergent. The hilum is large, 1½ mm. long, concave and depressed below a thick tan-colored rim. The chalaza is a prominent dark brown elevation.

Seeds of this species may occur incidentally with such crops as rice.

DESMODIUM

Desmodium tortuosum, Florida beggarweed (pl. XXXVII, 292).—Seeds are oval or ovate, flattened, glossy light reddish brown, 1½–2 mm. long, 3 mm. wide. The radicle is about half the length of the cotyledons and not divergent. The hilum is minute and depressed below a light-colored collar-like rim.

This species is in limited cultivation in the Southern States.

GALEGA

Galega officinalis, galega (pl. XXXVII, 293).—Seeds are oblong, much wider than long, length 1½–2 mm., width 4 mm. and thickness 1½ mm. Color is a dull dark reddish brown or olive green. The radicle is about half the length of the cotyledons or less, the tip not divergent. There is usually a pronounced depression on the surface of the seed between the radicle and the cotyledons. The hilum is minute, concave but lacks a distinct rim.

Seeds of galega may occur incidentally in imported seed of red clover and alfalfa from Italy. Compare this seed with *Trigonella* (pl. XL, 364).

GLYCINE

Glycine max, soybean (pl. XXXVII, 294).—There are well over 100 varieties of soybeans in commerce. Because new varieties are being introduced rather frequently, sometimes replacing older varieties, the overall picture on a countrywide basis is subject to continual change.

It is not practical to treat the identification of varieties of such a large group adequately within the scope of this paper.

The identification of the varieties by seed characters alone has certain possibilities and limitations. However, some features are pointed out which may serve as a guide to the agronomist or seed analyst who is concerned with only a few local varieties.

Diagnostic Features of Soybeans

Color

Three color groups are represented in soybean varieties.

Seeds black.—The number of varieties of black soybeans is relatively small, and not all of them are distinguishable. Three common varieties may be distinguished as follows:

1a. Seeds large, about 5½–6 mm. long and 9 mm. wide; oblong or oval and only slightly flattened .... *Glycine max* var. Wilson.
1b. Seeds smaller, about 5–6 mm. long and 7 mm. wide.
  2a. Oval and rotund .... *Glycine max* var. Otootan.
  2b. Oblong or oval and markedly flattened, faintly mottled with brown. *Glycine max* var. Laredo.

Seeds brown.—The number of varieties in this color group is also small and not all of them are distinguishable. Four common varieties are described.

1a. Seeds large, about 8 mm. long and 9 mm. wide; broadly oval and rotund; dark reddish brown .... *Glycine max* var. Biloxi.
1b. Seeds smaller, 5½–6 mm. long and 7½–8 mm. wide; oblong or oval.
  2a. Hilums with distinct whitish margins.
    3a. Seeds flattened, light reddish brown .... *Glycine max* var. Tanner.
    3b. Seeds rotund, dark reddish brown .... *Glycine max* var. Gatan.
  2b. Hilums without distinct whitish margins; seeds flattened, dull grayish brown .... *Glycine max* var. Virginia.

Yellowish tans and greens.—The yellow and green varieties comprise a vast group. The colors range from a pale creamy color, light to dark tan, and light green fading to yellowish green or greenish tan. A few varieties are a bright green. These colors may vary in degree due to conditions of development or age of seed. Consequently, color in this group is not in itself a reliable criterion and other features must be considered.

Hilum

Color of the hilum in the yellow-green group is an important diagnostic feature. The hilum in some varieties is deep black, with either a wide or a narrow band of black, gray or brown around its margins. The varieties Clark and Hawkeye may serve as an illustration. A light to dark reddish brown hilum, with a band of like color around its margin, is characteristic of another and larger group of varieties. The varieties Wabash and JEW–45 show this type of hilum. A third group has hilums that are very pale or colorless or with a pinkish cast, and the colored band around the margins is lacking. This type of hilum is illustrated in the varieties Gibson and Goldsoy.

Shape

Shapes of seeds should be evaluated cautiously. Well-developed seeds of some varieties have a characteristic shape, such as circular and rotund, oval or oblong and flattened. When grown under unfavorable conditions the seeds may not develop their true form. For example, plants from a seed that is normally round and rotund may produce many seeds that are oval and more or less flattened.

Size

As in other kinds of seeds, size is useful only in the extreme ranges that is, a seed may fall within the limits of the very small-seeded group, such as the variety Gibson, or the very large-seeded group such as the variety Mandarin. Between the small- and the large-seeded groups there is an intermediate group in which size is too indefinite for diagnostic purposes.

Surface

Well-developed seeds of some varieties are lustrous or oily, other varieties may be dull. This feature may be of value only when a small number of varieties are involved.

HEDYSARUM

Hedysarum coronarium, sulla (pl. XXXVII, 295).—Seeds are oval or roundish in outline. Color is yellow or light reddish brown and highly glossy. The radicle is about half the length of the cotyledons, the tip pointed and sharply divergent, the minute hilum lying within the sharp angle. Overall length and width are about equal, 2–3 mm., and the seed is thickest through the middle portion.

The species has been grown experimentally in the United States. The seeds may occur incidentally with crop seeds imported from Southern Europe.

INDIGOFERA

Indigofera hirsuta, hairy indigo (pl. XXXVII, 297).—Seeds are shaped like miniature satchels, the minute hilum lying where a handle would be placed. The length and width are about equal, 1½–2 mm. Color is pale yellowish or greenish with prominent dark green pits. The position of the radicle is not evident on the surface of the seed.

Hairy indigo is in limited use as a pasture plant in some coastal areas of the Southern States.

LATHYRUS

The genus *Lathyrus* has few species of agricultural importance. Nine species are described. Three of the species are grown as minor crops: *L. hirsutus*, *L. sylvestris*, and *L. tingitanus*. Six species may occur as contaminants in crop seeds such as vetch and the grains. Two species occur as incidental weed seed in imported seed: *L. annuus* and *L. stipularis*.

Similar-Appearing Species

L. hirsutus, roughpea (pl. XXXVII, 299).—Seeds of roughpea are quite similar to those of *L. tuberosus*, groundnutpea (pl. XXXVII,

302). Well-developed seeds appear to be distinguishable but some seeds are doubtfully distinct.

L. annuus (not included in the seed key) is another species that is similar to *L. hirsutus*. Seeds of this species may be found in imported vetches. The seeds tend to be larger and more spherical in shape than those of roughpea. The clear-cut protuberances on the seed coat are cone-shaped, the bases touching. The hilum is depressed, with a distinct rim around it.

L. angulatus, annual vetchling and *L. pusillus*, low vetchling.—Although the seeds of these two species are distinctive in shape they are sometimes confused. *L. pusillus* occurs chiefly in Eastern and Southern United States, and *L. angulatus* is established locally in the Pacific Northwest.

Key to *Lathyrus*

1a. Seed surface smooth.
- *2a.* Seeds glossy, broadly oval and flattened; length 4 mm. or more, width 3 mm.; reddish brown or pale greenish brown flecked with black, or almost entirely black; hilum oval, length ¾ mm., width ½ mm.
  L. aphaca, yellow vetchling (pl. XXXVII, 298).
- *2b.* Seeds not glossy.
  - *3a.* Length 2½ mm., width 2 mm.; variously compressed; buff color flecked with black or almost entirely black; hilum oval, length 1 mm., width ½ mm. ---------- *L. stipularis*, slender vetchling.
  - *3b.* Length 4–6 mm.; chalaza a large dark spot near the hilum.
    - *4a.* Seed roughly spherical, diameter about 4 mm.; reddish brown or greenish and minutely spotted with black, sometimes obscurely so; hilum oval, length 1 mm., width ½ mm.
      L. sphericus, round vetchling.
    - *4b.* Seed broadly oblong, flattened, length 5½–6 mm., width 7–8 mm.; dark brown, obscurely streaked or flecked with black, hilum oblong with a ridge around the margin, length 3½–4 mm., width 1 mm.
      L. tingitanus, Tangier pea (pl. XXXVII, 301).

1b. Seed surface not smooth.
- *5a.* Hilum linear, about ⅓ the length of the circumference, width ½ mm. Seed spherical, diameter 4½–5 mm.; reddish brown, faintly roughened by irregular ridges ---------- *L. sylvestris*, flatpea (pl. XXXVII, 300).
- *5b.* Hilum oval, 2 mm. long or less, 1–1½ mm. wide.
  - *6a.* Seed spherical or compressed at the ends in *hirsutus*.
    - *7a.* Length 1½–2 mm.; light reddish brown with numerous black flecks; surface minutely stippled and with a network of high, thin elevations; hilum with a high, thin rim around the margin.
      L. pusillus, low vetchling.
    - *7b.* Length 3–3½ mm.; surface roughened by close-spaced, knob-like tubercles, or ridges; chalaza a large, smooth spot near the hilum.
      L. hirsutus, roughpea (pl. XXXVII, 299).
  - *6b.* Seeds not spherical.
    - *8a.* Seeds oval or oblong, length 3 mm., width 4½–5 mm.; surface very minutely stippled and roughened by flattened ridges or tubercle-like projections ---------- *L. tuberosus*, groundnutpea (pl. XXXVII, 302).
    - *8b.* Seeds 4-sided and thick, the end planes somewhat triangular, the small, oval hilum in one of the wider angles on one end; reddish or grayish brown, finely tubercled; length 2–2½ mm., width 2 mm.
      L. angulatus, annual vetchling.

LESPEDEZA

Three species of lespedeza are commercially important in the United States: *L. striata*, *L. stipulacea*, and *L. cuneata*. Two species are in more limited use in the Southern States for game food and erosion control: *L. bicolor* and *L. thunbergii*. *L. hedysaroides* is of experimental interest. There are numerous native species and the seeds of

600705 O—62——6

some of these may occur with crop seeds. Two such species are included here *L. stuevei* and *L. capitata*.

Fruits of lespedeza are 1-seeded and consist of a pod, calyx, and three scale-like bracts. Commercial seed usually consists of the entire fruits with a few seeds that have been hulled in processing. Some kinds, such as *L. cuneata*, may consist entirely of hulled seeds. The calyx and bracts in some species are removed in processing and the seed unit is the pod only, with seed.

Features Common to All Species

Shape and pubescence of pods.—Plants of some species normally produce two types of flowers. It has been reported that pods from petaliferous flowers are more pointed than those from apetalous flowers. Space does not permit showing these different types in the illustrations. Pubescence, except as noted in the keys, is whitish, long, and appressed.

Color of pod and calyx.—The species under consideration vary in color from pale brownish to reddish or grayish brown. The calyx is strongly 3-nerved. Venation of pods is a network or reticulum of dark brown lines that may be faint or prominent in different species.

Hulled seeds.—Hulled seeds of the species included here are glossy. The hilum is minute and sunken with a whitish collar-like margin. Shape of seeds in most species is ovate. The hilum, with one exception, lies on one edge of the long axis of the seed about midway or higher. *L. stipulacea* has more uniformly oval seeds and the hilum lies at one end or near it and is scarcely indented.

Similar-Appearing Species and Varieties

Lespedeza striata, varieties Common and Kobe.—There is a marked difference in size, both in pods and hulled seeds, when viewed in bulk. The difference may not always be apparent when individual seeds are viewed. With experience, it is usually possible to recognize the reddish color aspect of the pod and calyx of Common lespedeza as distinguished from the grayish-brown color of the variety Kobe. The densely ciliate margins of the calyx lobes are strongly indicative of Common lespedeza but these hairs are sometimes damaged in processing.

Hulled seeds.—The species may be placed into three color groups: (1) Seeds light green or green flecked with purple, (2) purple with light-colored flecks, and (3) seeds entirely purple. Seeds of the species within each color group are distinguished mainly by size. As in other seeds, size is useful mainly in the extreme ranges, that is, a seed may fall within the range of a small-seeded group or a very large-seeded group. Because of natural variation there may be an overlapping in size between species, and individual seeds may not always be identified with certainty.

SEED KEYS

Since commercial seed may contain seeds both with and without the pod, two seed keys are provided. One key describes the seeds in the pod, the other key describes hulled seeds.

Key to Pods of *Lespedeza*

1a. Calyx slit to near the base, the lobes narrow and acuminate, pubescent; pods pubescent across the top and along the sides, the reticulum faint; length 3–5 mm., width 1¾–2 mm.

2a. Pods dark reddish brown; calyx about ¾ the length of the pod.
L. cuneata, sericea lespedeza (pl. XXXVII, 306).

2b. Pods grayish brown; calyx equal to or longer than the pod.
L. hedysaroides, Siberian lespedeza.

1b. Calyx not slit to the base.

3a. Calyx about ¾ the length of the pod, the lobes broad and obtuse; pods sparsely pubescent, the reticulum faint.

4a. Pod and calyx reddish brown, the calyx densely pubescent along the margins of the lobes; bracts narrow, with prominent nerves; length 3 mm., width 2 mm.----*L. striata*, Common striate lespedeza (pl. XXXVII, 303).

4b. Pod and calyx grayish brown, the calyx sparsely pubescent; bracts wider and nerves more numerous than above; length 4–5 mm., width 2½ mm.
L. striata, Kobe striate lespedeza (pl. XXXVII, 304).

3b. Calyx about one-half the length of the pod or less.

5a. Calyx lobes broad and obtuse, glabrous; pods broadly oval, sparsely short-pubescent, reticulum prominent; length 3–3½ mm., width 2½ mm.
L. stipulacea, Korean lespedeza, including the vars. Climax and Rowan. (pl. XXXVII, 305).

5b. Calyx lobes acute, sparsely pubescent; pods copiously long-pubescent, the reticulum faint; length 6–7 mm. or longer, width 4–5 mm.
L. bicolor, bicolor lespedeza.

Key to Hulled Seeds of *Lespedeza*

1a. Seeds light green, green flecked with purple or brownish purple; shape ovate, the hilum on one side about midway of the long axis of the seed or a little higher.

2a. Length 2–2½ mm., width 1½ mm.
L. cuneata, sericea lespedeza (pl. XXXVII, 306).
L. hedysaroides, Siberian lespedeza.

2b. Length 3 mm., width 2 mm.------------*L. capitata*, roundhead lespedeza.
L. stuevei, Stueve lespedeza.

1b. Seeds not green or green flecked with purple.

3a. Color uniformly purple.

4a. Seeds broadly oval with a faint line between the radicle and cotyledons; hilum at or near one end, scarcely indented, the whitish hilar margin not prominent; length 2½–3 mm., width 2–2½ mm.
L. stipulacea, Korean lespedeza, including the vars. Climax and Rowan. (pl. XXXVII, 305).

4b. Seeds broadly ovate or oval; hilum on one side near the middle of the long axis of the seed, the whitish hilar margin prominent; length 3½–4 mm., width 2½–3 mm.------------------*L. thunbergii*, Thunberg lespedeza.

3b. Color not uniformly purple.

5a. Color purple with light-colored flecks, occasional seeds not flecked; shape ovate, the hilum on one side about midway of the long axis of the seed or a little higher, the hilar margins not prominent.

6a. Length 2 mm., width 1½ mm.
L. striata, Common striate lespedeza (pl. XXXVII, 303).

6b. Length 2½ mm., width 2 mm.
L. striata, Kobe striate lespedeza (pl. XXXVII, 304).

5b. Color pale brownish, copiously flecked with purple; shape broadly ovate, the hilum on one side about midway of the long axis of the seed, the hilar margin prominent; length 3 mm., width 2½ mm.
L. bicolor, bicolor lespedeza.

LOTUS

Two species of *Lotus* are in limited cultivation in the United States: *L. corniculatus*, birdsfoot trefoil and *L. uliginosus*, big trefoil.

There are two botanical varieties of *L. corniculatus*. The variety *arvensis* has broad leaflets, the variety *tenuifolius* has narrow leaflets. The two varieties are considered by some authors as distinct species.

Lotus uliginosus also has two botanical varieties. The more widely used variety, *villosus*, is a pubescent form; the variety *glabriusculus* is glabrous or nearly so. Importations of *L. uliginosus* are sometimes a mixture of the two varieties.

Three species are of experimental interest and the seeds may occur as incidental seeds with crop seeds: *L. purshianus* (*L. americanus*), prairie trefoil; *L. angustissimus*, slenderpod deervetch; and *L. hispidus*, hispid deervetch.

Key to *Lotus*

1a. Hilum surrounded by a prominent whitish collar-like structure; seeds broadly oval and rotund.
  2a. Color dull brown, often obscurely flecked with purple.
    L. corniculatus, birdsfoot trefoil (pl. XXXVII, 308).
  2b. Color olive-green, brownish or reddish brown, not flecked or mottled.
    L. angustissimus, slenderpod deervetch (pl. XXXVII, 307).
1b. Hilum not surrounded by a prominent whitish collar.
  3a. Seeds oblong and slightly flattened, length 2 mm., width 3 mm.; color greenish or reddish brown, mottled with darker green and flecked with black; hilum in a shallow indentation about midway of the long axis of the seed.
    L. purshianus, prairie trefoil (pl. XXXVIII, 310).
  3b. Seeds broadly oval and rotund, length and width equal, 1–1¼ mm.
    4a. Seeds not mottled or flecked; bright yellow-green or brownish.
      L. uliginosus var. *glabriusculus*, smooth big trefoil (pl. XXXVIII, 311).
    4b. Seeds mottled or flecked with purple.
      5a. Color dull greenish or brownish, copiously mottled.
        L. uliginosus var. *villosus*, hairy big trefoil (pl. XXXVIII, 312).
      5b. Color dull brown or purplish, sparingly flecked with dark purple.
        L. hispidus, hispid deervetch (pl. XXXVII, 309).

LUPINUS

Three species of *Lupinus* are of commercial importance in the United States: *L. angustifolius*, blue lupine; *L. luteus*, yellow lupine; and *L. albus*, white lupine. Several horticultural varieties of these species are in production, the most important group being the nonalkaloid varieties of blue lupine.

Seeds of lupine are broadly rectangular and more or less flattened. The hilum lies across one of the obtuse angles.

Seeds of the alkaloid varieties of blue lupine are pale gray, those of the nonalkaloid varieties are grayish brown. When viewed in bulk, the darker, brownish color of a nonalkaloid variety distinguishes it from the alkaloid varieties. It would be difficult, however, to distinguish all seeds in a mixture with certainty.

Key to *Lupinus*

1a. Hilum about 1 mm. long; seeds dull.
  2a. Seeds flattened; color creamy white, streaked and flecked with black, or entirely white in some varieties; length 9 mm., width 7 mm.
    Lupinus luteus, yellow lupine (pl. XXXVIII, 314).
  2b. Seeds only slightly flattened; color dull gray or brownish with minute white spots and obscure brown flecks; length 8–9 mm., width 6 mm.
    Lupinus angustifolius, blue lupine (pl. XXXVIII, 313).
1b. Hilum about 2 mm. long; seeds flattened, creamy white with pinkish cast and semipolished; length 10–15 mm., width 10–12 mm.
    Lupinus albus, white lupine.

MEDICAGO AND *MELILOTUS*

For structural features of legume seeds refer to figs. 12 and 13, p. 66.

Seeds of some species of *Medicago* and *Melilotus* are very similar, and for this reason the two genera are considered together and included in one seed key.

Six species of the genus *Medicago* are of economic interest. Alfalfa (*M. sativa*), with several horticultural varieties, is the most widely grown. Spotted bur-clover (*M. arabica*) and California bur-clover (*M. hispida*) are in more limited production in the Southeast and West, respectively. A form of *M. hispida* has spineless burs. Black medic (*M. lupulina*) is not an important crop in the United States, although the seed is available commercially. The seed occurs quite commonly with other agricultural seeds. Buttonclover (*M. orbicularis*) is a minor crop in some Southern States. Cogwheel clover (*M. tuberculata*) is of experimental interest at the present time. Buttonclover and the bur-clovers may occur incidentally with such crop seeds as alfalfa and red clover, especially from Italy and France.

Three species of *Melilotus* are in cultivation in the United States. White sweetclover (*M. alba*) and yellow sweetclover (*M. officinalis*) are crops of major importance throughout the corn belt and northward into Canada. Several horticultural varieties of these two species, both annual and biennial, are in commerce. Sourclover (*M. indica*) is in limited production in the South and Southwest. Seeds of any of the three species may occur incidentally with other seeds from these general areas.

Similar-Appearing Species

Alfalfa and sweetclover.—Alfalfa often contains a natural admixture of seeds of sweetclover which are rather difficult to distinguish. Because of the numerous horticultural varieties involved, color and texture are not reliable criteria and the structural features of the seeds must be carefully evaluated.

The coiled pods of alfalfa produce seeds that are variously compressed and misshapen. As pointed out in the key that follows, the seeds may fall into three categories based on shape and position of the hilum (pl. XXXVIII, 319). The oval type with the hilum near one end is the form most likely to be confused with sweetclover.

Identification of questionable seeds may be made by examining them in edge view with the radicle edge uppermost. In this position one can see that the mass of the alfalfa seed is not evenly distributed. Both sides of the seed show irregular bulges and appear lumpy, with a slight twist or bend in the long axis of the seed.

Seeds of sweetclover, when viewed in the same edgewise position, will lack the bend or twist and the lumpiness that is characteristic of alfalfa. One side of the seed is flat, the other side is slightly rounded, and the thickness is quite uniform throughout.

Alfalfa and California bur-clover (*M. hispida*).—Seeds of California bur-clover may be difficult to detect in a sample of alfalfa. There is an obvious difference in the appearance of the chalaza. The chalaza of the alfalfa seed is relatively inconspicuous and that of bur-clover is a prominent brown spot. Cotyledons of bur-clover are bulged out adjacent to the radicle, but the seed in edge view is not lumpy or bent as in alfalfa.

Sweetclover.—When in the pods, seeds of white-blossom and yellow-blossom sweetclover are distinguishable. (See pl. XXXVIII, 321 and 322.) Hulled seeds are generally indistinguishable except those seeds of yellow-blossom that show purple mottling. The purple spots may be so faint on some seeds that they can be detected only under magnification.

The mottled seeds present in a sample of yellow-blossom sweetclover may vary from only a trace to almost the entire sample consisting of mottled seeds. This depends somewhat on the variety. There is evidence also that the extent of mottling may be influenced by climatic factors.

Key to *Medicago* and *Melilotus*

1a. Seed coat rough; shape broadly oval.
  2a. Surface roughened by minute tubercles (seeds grate when rubbed together); color dark olive-green; radicle shorter than the cotyledons, the hilum in shallow indentation on one side above the middle; length 2 mm., width 1½ mm. ---------------- *Melilotus indica*, sourclover (pl. XXXVIII, 323).
  2b. Surface with blister-like protuberances; color tawny to reddish-brown; radicle equal to cotyledons or nearly so, forms a straight edge along one side of seed, the tip divergent, a pronounced depression between the radicle and cotyledons; hilum in a sharp notch at one end of the seed; length 2½–3 mm., width 3 mm.
    Medicago orbicularis, buttonclover (pl. XXXVIII, 318).
1b. Seed coat smooth.
  3a. Seeds kidney-shaped; radicle about half the length of the cotyledons, hilum near the middle of the long axis of the seed; pods coiled, several-seeded.
    4a. Tip of radicle divergent, forming a distinct point at the hilum.
      5a. Seeds flat, thin, both faces plane; color light yellow or brownish; length 1½ mm., width 3 mm., spines of pod fine, stiff, reflexed.
        Medicago arabica, spotted bur-clover (pl. XXXVIII, 315).
      5b. Seeds not thin and flat, in edge view both faces appear lumpy with a slight bend or twist in the long axis of the seed; length 1½ mm. or more, width 2½–3 mm.; pods spineless.
        Medicago sativa, alfalfa (in part) (pl. XXXVIII, 319).
    4b. Tip of radicle not divergent, the hilum in a broad and shallow indentation; seeds not thin and flat.
      6a. Chalaza a prominent brown spot near the top of the cotyledons; a distinct groove between radicle and cotyledons; cotyledons bulged out adjacent to the radicle but not lumpy and bent as in *sativa*; length 1½ mm. or more, width 3 mm.; spines of pod slender, rigid and upright, usually hooked at the tip.
        Medicago hispida, California bur-clover (pl. XXXVIII, 316).
      6b. Chalaza not prominent.
        7a. Radicle forms a straight line; cotyledons smoothly rounded, often with an imprint of the coiled pod on the back edge; length 1½ mm., width 4½–5 mm.; pods spineless, with prominent configurations on the back.
          Medicago tuberculata, cogwheel-clover (pl. XXXVIII, 320).
        7b. Radicle not a straight line; cotyledons appear lumpy and with a slight bend or twist in the long axis of the seed; length 1½ mm. or more, width 2½–3 mm.
          Medicago sativa, alfalfa (in part) (pl. XXXVIII, 319).
  3b. Seeds not kidney-shaped; pods 1-seeded (exception in *sativa*).
    8a. Seeds ovate, both sides smoothly rounded; radicle divergent at the tip, forming a point at the hilum near the narrower end of the seed; length 1-1½ mm., width 2 mm. or more; pods black, the nerves in concentric lines ------------ *Medicago lupulina*, black medic (pl. XXXVIII, 317).
    8b. Seeds oval; radicle slightly shorter than the cotyledons, the tip not divergent, the hilum in a shallow indentation near one end; a conspicuous

whitish stripe between the radicle and cotyledons; length 2½ mm., width 1½ mm. or more, but may vary with the variety.

9a. One side plane (seed lies flat on the table), the other side smoothly rounded.

10a. Seeds yellowish or greenish yellow, not mottled with purple; pod with a network of coarse nerves, the calyx not inflated below the pod--------*Melilotus alba*, white sweetclover (pl. XXXVIII, 321).

10b. A variable proportion of seeds mottled with purple; pod with coarse nerves lying crosswise of the pod, the calyx inflated below the pod.
Melilotus officinalis, yellow sweetclover (pl. XXXVIII, 322).

9b. Seeds with both sides lumpy and with a slight bend or twist in the long axis of the seed; does not lie flat on the table.
Medicago sativa, alfalfa (in part) (pl. XXXVIII, 319).

ONOBRYCHIS

Onobrychis viciaefolia, sainfoin (pl. XXXVIII, 324).—The seed unit is usually the indehiscent pod, 6–7 mm. long. The sides of the pod are coarsely reticulate; the curved edge is margined and coarsely toothed toward the top.

Hulled seeds are about 3 mm. long, 4½–5 mm. wide, semicircular in shape, dark reddish brown in color. The hilum lies in a shallow indentation near the middle of the straight edge of the seed.

Although adaptable throughout the alfalfa regions of this country, it is only in very limited production. The seeds sometimes occur incidentally in imported crop seeds.

ONONIS

Ononis repens, ononis (pl. XXXVIII, 325).—The seeds are flattened and roughly orbicular in shape; the length and width are about equal, 2½–2¾ mm.; the color is dull greenish or reddish brown; the hilum is minute, concave, depressed below the seed surface, and usually lies in a shallow indentation. Compare with the somewhat similar seeds of *Medicago orbicularis*, (pl. XXXVIII, 318). Seeds of ononis may occur in imported clover and alfalfa.

ORNITHOPUS

Ornithopus sativus, serradella (pl. XXXVIII, 326).—The pods consist of several indehiscent jointed segments with anastamosing raised lines. The segments are 3–4 mm. long and 2–3 mm. wide, with a smooth callus across each end. Hulled seeds are not likely to be found in seed samples.

The species is not a cultivated crop in this country. The indehiscent segments of the pod may occur incidentally in imported crop seeds.

PHASEOLUS

Seeds of different species of *Phaseolus* vary in size and color but certain features are common to all of them. The shape is mostly oblong or oval and slightly flattened. Hilums lie on the long axis of the seed and tend to be narrow in proportion to their length, the surface obscured by a layer of whitish corky tissue.

Seeds of *Phaseolus* are somewhat similar to those of cowpeas (*Vigna sinensis*). For a comparison of the seeds of these two genera, refer to *Vigna*, pp. 92-93.

Phaseolus vulgaris, field and garden beans.—This group is not included here. The main problem involves the identification of the many horticultural varieties and such a large group could not be treated adequately within the scope of this paper.

Phaseolus aureus, mung bean, also known as green or golden gram (pl. XXXVIII, 328).—Seeds are oblong and slightly flattened or almost spherical, 3½ mm. long, 4–4½ mm. wide; color is light olive green with usually a band of darker green around the hilum. The hilum is ovate or oblong, 1½ mm. long and ½ mm. wide, obscured by a layer of whitish corky tissue, the layer elevated above the seed surface.

Mung beans are grown in the Southwest for human food and as a soil-improvement crop.

Phaseolus angularis, adzuki bean (pl. XXXVIII, 327).—Seeds are oblong, slightly flattened, 5 mm. long, 7–8 mm. wide, and a glossy mahogany color. The hilum is 3½–4 mm. long, ½–¾ mm. wide, obscured by a layer of whitish corky tissue which is flush with the seed surface or nearly so.

Adzuki beans are grown to some extent for home use.

PISUM

Pisum sativum var. *arvense*, field pea (pl. XXXIX, 329).—There are many horticultural varieties of field pea, but the variety Austrian Winter is planted most extensively in the United States. The seeds are spherical, about 6 mm. in diameter, grayish to reddish brown and minutely spotted with black, and frequently mottled with darker brown. Hilum is oval or ovate, 1½–2 mm. long, flush with the seed surface and usually scurfy. The chalaza is about 2½ mm. from the hilum.

PUERARIA

Pueraria thunbergiana, kudzu (pl. XXXIX, 330).—Seeds of kudzu are almost semicircular in outline, 3 mm. long or less and 4–5 mm. wide. The color is creamy white or light reddish brown, variously streaked and mottled with black. Occasional seeds are light reddish brown without mottling. The hilum lies about midway on the long axis of the seed and has an upstanding collar of whitish corky tissue around its margin.

Kudzu is adapted to the Southeastern States where it is useful for hay, soil improvement, and erosion control. It is reported to be a poor seeder and the seed is not in commercial production.

SESBANIA

Sesbania exaltata, sesbania (pl. XXXIX, 331).—Seeds are oblong, 2–2½ mm. long and 4 mm. wide, slightly flattened, the thickness uniform throughout. Color is greenish or light brown, variously streaked and mottled with black. The hilum is minute, concave, with a narrow whitish rim around the margin.

Sesbania is planted as a soil-improvement crop in irrigated sections of the Southwest and the seed may occur as an impurity in such crops as rice.

STIZOLOBIUM

Stizolobium deeringianum, velvetbean (pl. XXXIX, 332).—Seeds are oval, creamy white or gray copiously mottled with black, 10 mm. long, 14–15 mm. wide. The hilum lies on the long axis of the seed. It is about 5 mm. long and 1 mm. or more wide, surrounded by a thick whitish collar with crimped margins. The seeds are readily recognized by the large size and the characteristically ornamented collar around the hilum.

Velvetbean is adapted to the southern part of the United States. It is not an important agricultural crop at the present time.

STROPHOSTYLES

Strophostyles leiosperma, smooth-seeded wildbean (pl. XXXIX, 333).—Seeds are oblong and slightly flattened, 2½ mm. long and 3 mm. wide, glossy, gray and densely spotted with black so that seed appears almost black. The hilum, about 2 mm. long, is obscured by a thick layer of whitish corky tissue. The seed coat is elevated into a ridge around the hilum, and the chalaza is prominent and knob-like.

This seed may occur incidentally with lespedeza or similar crops.

TRIFOLIUM

When considered on a countrywide basis, the true clovers (*Trifolium*) comprise a very large group. Thirty-two species are included here. About nine of the species are of major agricultural importance in the United States. These are indicated in the seed key by an asterisk (*). Seven species are crops that are restricted to more local conditions and may be considered as minor crops. These are indicated by two asterisks (**). Some of the remaining species are being grown experimentally or occur as incidental weed seeds with crop seeds.

Structural Features

Seeds of some of the clovers are quite similar. In order to distinguish them the morphological structures must be relied upon rather than the general appearance of the seed. Before using the seed key it would be helpful to review the structure of legume seeds on page 66, figs. 12 and 13.

Shape.—The radicle is bent back against the edge of the cotyledons. The shape of the seed is determined largely by the relative lengths of the radicle and cotyledons, the character of the radicle and the position in which it lies against the cotyledons (fig. 13, p. 66). This in turn determines the position of the hilum which lies near the tip of the radicle.

Size.—Because the hilum lies in different positions on the seed, the lengths stated in the key represent the overall length of the seed from tip to base. Width represents the widest point of the seed. All measurements should be considered approximate only.

Surface and color.—Comparatively few of the species have a roughened surface. Sometimes the tubercles are minute and a magnification of 10 × or 20 × may be required to detect them. These seeds are usually dull.

The smooth seeds may be highly glossy or dull. Seeds that normally are rather dull may become polished in processing and appear slightly lustrous. It is usually more satisfactory to examine the smooth seeds at a low magnification of about 6 × rather than a higher magnification.

Similar-Appearing Species

The seed key gives quite detailed descriptions of each species. In addition, attention is called here to those species that are superficially very similar and may require a more careful comparison of the distinguishing features. A few species appear in more than one place in the seed key, mainly because of color.

T. dubium, suckling clover (pl. XXXIX, 342) and *T. campestre* (*T. procumbens*), large hop clover (pl. XXXIX, 343).—Suckling clover may vary greatly in size in different seed lots. Large hop clover is uniformly smaller and flatter than suckling clover. In addition to the features listed in the seed key, it has been observed that the radicle in suckling clover bulges out below the tip while that of large hop clover is nearly straight.

T. incarnatum, crimson clover (pl. XL, 361) and *T. striatum*, striate clover (pl. XL, 350).—Striate clover is a minor crop in the South. The seed sometimes occurs in samples of crimson clover, especially in imported seed. The two species are somewhat similar but they are readily distinguished.

T. nigrescens, ball clover (pl. XXXIX, 346).—Ball clover often carries an admixture of *T. repens*, white clover (pl. XXXIX, 348). White clover is quite variable in shape and could be confused with some seeds of ball clover. Ball clover is smaller in size and tends to be variously warped or misshapen.

T. michelianum, big-flowered clover (pl. XXXIX, 345) and *T. resupinatum*, Persian clover (pl. XL, 349).—It will be observed in the illustrations that the radicle in both species is about equal to the cotyledons. As shown in the illustrations, the most outstanding difference between the two species is the shape of the seed at the end opposite the hilum.

Other similar species.—Without the calyx or involucre present, seeds of some species may be indistinguishable. This applies to the following species: *T. gracilentum* (pl. XL, 353) and *T. ciliolatum; T. bifidum* (pl. XL, 351), *T. microdon* (pl. XL, 356) and *T. microcephalum* (pl. XL, 355).

Key to *Trifolium*

The seed key is necessarily long and detailed in order to include the large number of species. The key is arranged in four sections as follows:

Section 1.—Seeds roughened by tubercles or wrinkles (p. 87).
Section 2.—Seeds smooth, color yellow (pp. 87-88).
Section 3.—Seeds smooth, spotted or streaked with purple, dark green or black (p. 89).
Section 4.—Seeds smooth, color olive green, purple (at least in part), or black (pp. 89-90).

Key to *Trifolium*

Section 1

Surface roughened by minute tubercles or coarse wrinkles; seeds dull.

1a. Seeds small (½–¾ mm.), length and width about equal; yellow or occasional seed bluish green in *glomeratum;* oval, the radicle equal to or slightly shorter than the cotyledons, the hilum on or near one end.

2a. Rotund, the radicle not prominent; tubercles minute, the projections not evident around the periphery.
****T. glomeratum*, cluster clover (pl. XXXIX, 335).

2b. Flattened, the radicle thick and prominent; tubercles coarser, the projections evident around the periphery.
T. parviflorum, teasel clover (pl. XXXIX, 338).

1b. Seeds markedly larger (1¼–2 mm. long), light olive green or brownish, some seeds lightly streaked with purple in *reflexum* (*4b*); broadly oval or roundish, the radicle thick, equal to or exceeding the cotyledons.

3a. Surface coarsely wrinkled or with wart-like tubercles; hilum in a deep notch at one end; length 2 mm., width 1½ mm.
T. depauperatum, poverty clover (pl. XXXIX, 339).

3b. Surface finely roughened by flattened tubercles or short irregular lines; length and width about equal, 1¼–1½ mm.

4a. Radicle tip widely divergent, a deep crease between radicle and cotyledons; hilum in a deep notch on one end.
T. carolinianum, Carolina clover (pl. XXXIX, 337).

4b. Radicle tip not widely divergent, the crease between radicle and cotyledons faint or lacking; hilum in a shallow indentation on one end.
T. reflexum, buffalo clover (pl. XXXIX, 336).

Key to *Trifolium*

Section 2

Surface smooth, color yellow. (Some exceptions in *dubium*, *3a*, and *agrarium*, *8a*.)

1a. Seeds highly glossy.

2a. Seeds small, 1½ mm. long or less, ¾–1 mm. wide (some strains of *dubium* larger); radicle inconspicuous, slightly shorter than the cotyledons, the tip not divergent; hilum scarcely indented.

3a. Broadly oval, rotund, the sides strongly curved, in edge view usually thickest from the middle to the hilar end; many seeds purplish.
**T. dubium*, suckling clover (pl. XXXIX, 342).

3b. Narrowly oval or oblong, flattened, the sides scarcely curved; in edge view the thickness about uniform throughout or slightly thicker in lower half.
T. campestre (*T. procumbens*), large hop clover (pl. XXXIX, 343).

2b. Seeds larger, up to 3 mm. long and 2 mm. wide; commonly with a greenish tinge in *tridentatum* (*5a*).

4a. Radicle about ½ the length of the cotyledons, inconspicuous; seed oval, rotund, the sides smoothly rounded; hilum on the side, scarcely indented; a thick rim around the depressed hilar area.
**T. incarnatum*, crimson clover (pl. XL, 361).

4b. Radicle equal to or exceeding the cotyledons; hilum on one end or nearly so.

5a. Broadly oval or oblong, the ends broad and blunt; one face flat, the other humped up about midway; radicle thick, the hilum in a sharp notch; length 2–2½ mm., width 1¾–2 mm. (See also sec. 3, *3b*, p. 89.)
T. tridentatum, tomcat clover (pl. XL, 357).

5b. Ovate, smoothly rounded; hilum in a shallow indentation on the broader end, the opposite end pointed; radicle inconspicuous; some seeds olive green. (See sec. 4, *2a*, p. 89.)
**T. resupinatum*, Persian clover (pl. XL, 349).

1b. Seeds dull or lustrous.

6a. Seeds small, 1¼–1½ mm. long, ¾–1 mm. wide. (See also *12a.*)

7a. Uniformly oval, the sides smoothly rounded; radicle shorter than the cotyledons, the tip not divergent; hilum scarcely indented.

8a. Radicle distinct, usually with a fine yellowish line between it and the cotyledons; deep yellow toward the hilar end and green below: flattened.
T. agrarium, yellow hop clover (pl. XXXIX, 340).

8b. Radicle inconspicuous, the line or crease between it and the cotyledons scarcely evident; greenish yellow; rotund.
T. arvense, rabbitfoot clover (pl. XXXIX, 341).

7b. Asymmetrically oval or triangular, often warped or variously misshapen; radicle mostly markedly shorter than the cotyledons, the tip scarcely divergent, a distinct crease between it and the cotyledons; hilum in a slight indentation on the side. (Compare with *12a.*)
***T. nigrescens*, ball clover (pl. XXXIX, 346).

6b. Seeds larger, 1½–3 mm. long, 1¼–2 mm. wide.

9a. Seeds roundish; radicle inconspicuous, slightly shorter than the cotyledons; hilum scarcely indented.

10a. Length and width about equal, 1¼ mm. or more; rotund, with minute whitish spots evident beneath the surface.
***T. striatum*, striate clover (pl. XL, 350).

10b. Length and width about equal, 2–2½ mm. or more; flattened; pale yellow, the seed coat with a few short longitudinal wrinkles at the tip of the cotyledons---- **T. hirtum*, rose clover (pl. XL, 360).

9b. Seeds other than roundish and rotund.

11a. Broadly heart-shaped, triangular or oval; radicle thick, the tip strongly divergent.

12a. Radicle equal to the cotyledons or nearly so, the hilum between the two lobes on the end, or the radicle much shorter with the hilum on one side in a broad, shallow indentation; length and width approximately equal, 1½ mm.
**T. repens*, white clover (pl. XXXIX, 348).

12b. Radicle markedly shorter than the cotyledons; shape triangular; length 2–2½ mm.

13a. Width 1½–1¾ mm., cotyledon end narrow, the opposite end broadly rounded. (See also sec. 4, *4a*, p. 90.)
**T. pratense*, red clover (pl. XXXIX, 347).

13b. Width 2–2½ mm.; cotyledon end broad, the opposite end narrow and usually pointed. (Compare with *15b.*)
***T. medium*, zigzag clover (pl. XL, 362).

11b. Neither heart-shaped nor triangular; radicle tip not strongly divergent.

14a. Seeds 1½–1¾ mm. long, 1¼–1½ mm. wide; oval; radicle equal to the cotyledons or nearly so, a fine crease or white line between it and the cotyledons; hilum on one end or near it; color deep yellow, occasional seed purplish or dull olive green and yellowish around the hilar area. (Compare with *5b*, p. 87.)
***T. michelianum*, big-flowered clover (pl. XXXIX, 345).

14*b.* Seeds 2½–3 mm. long, 2 mm. wide or more; radicle shorter than the cotyledons.

15a. Broadly oval, a short, wide depression between radicle and cotyledons; hilum in an indentation or notch near one end, a thick rim around the depressed hilar area; some seeds streaked or spotted with dark purple or black. (See sec. 3, *8b*, p. 89.)
T. gracilentum, pinpoint clover (pl. XL, 353).
T. ciliolatum, tree clover.

15b. Ovate; hilum near the narrower end, slightly indented, the hilar area not depressed, the opposite end broadly rounded; a long, white inward curving line between the radicle and cotyledons. (Compare with *13b.*)
T. alexandrinum, berseem clover (pl. XL, 359).

Key to *Trifolium*

SECTION 3

Seeds smooth, spotted or streaked with purple, dark green or black.

1a. Seeds highly glossy.
  2a. Length 1½–1¾ mm., width 1¼ mm.; radicle equal to the cotyledons or nearly so; broadly oval or heart-shaped, the hilum between two lobes on one end; color yellowish and prominently or faintly spotted with black.
  T. ornithopodioides, birdfoot clover.
  2b. Length 2–3 mm., width 1½–2 mm.
    3a. Ovate, the radicle mostly shorter than the cotyledons, the hilum in a notch near the broader end, the opposite end narrower and often pointed; usually yellowish across the tip of the cotyledons, grayish below and prominently or faintly spotted with purple or black.
    T. variegatum, whitetip clover (pl. XL, 358).
    3b. Broadly oval or oblong, the ends broad and blunt, one side flat, the other side humped up about midway; radicle thick, equal to or exceeding the cotyledons, the hilum in a sharp notch; finely and copiously spotted with purple. (Compare with sec. 2, *5a*, p. 87.)
    T. tridentatum, tomcat clover (pl. XL, 357).

1b. Seeds dull or lustrous.
  4a. Broadly oval or heart-shaped.
    5a. Radicle equal to or exceeding the cotyledons; length 1½–1¾ mm., width 1¼–1½ mm.
      6a. Bright yellow or reddish and copiously spotted with black below the hilar area, with a short yellowish line between the radicle and cotyledons; hilum in a shallow indentation on the broader end.
      **T. fragiferum*, strawberry clover (pl. XL, 352).
      6b. Yellowish at the tip of the cotyledons, dull yellowish green below and copiously spotted with black, the line between radicle and cotyledons lacking; hilum in a notch on the broader end. (Compare with *3a*.)
      T. involucratum, seaside clover.
    5b. Radicle equal to or shorter than the cotyledons; hilum between two lobes on one end; length 1¼–1½ mm., width 1–1¼ mm.
      Light or dark bluish green, copiously spotted with darker green, some seeds almost black; immature seeds light bluish green.
      **T. hybridum*, alsike clover (pl. XL, 354).
  4b. Seeds neither broadly oval nor heart-shaped; radicle shorter than the cotyledons, the tip not divergent; color yellow or greenish and spotted or streaked with dark purple or black.
    7a. Narrowly ovate; hilum in a slight indentation near the narrower end, the opposite end broadly rounded; length 2–2½ mm., width 1¼ mm.
    T. bifidum, pinole clover (pl. XL, 351).
    7b. Oval, both ends about equally rounded; hilum usually in a distinct notch near one end.
      8a. Seeds slender.
        9a. Length 1½ mm., width 1 mm.
        T. microcephalum, small-headed clover (pl. XL, 355).
        9b. Length 2 mm., width 1¼ mm.
        T. microdon, thimble clover (pl. XL, 356).
      8b. Seeds broad, length 2½–3 mm., width 2 mm. Seeds with wide black streaks or entire surface spotted with dark purple or black, some seeds entirely yellow. (Compare with *15a*, sec. 2, p. 88.)
      T. gracilentum, pinpoint clover (pl. XL, 353).
      T. ciliolatum, tree clover.

Key to *Trifolium*

SECTION 4

Seeds smooth, color olive-green, purple (at least in part), or black.

1a. Seeds highly glossy.
  2a. Light to dark olive-green. (Compare with *5a*, sec. 2, p. 87, and *14a*, sec. 2, p. 88.) ---------------------- **T. resupinatum*, Persian clover (pl. XL, 349).
  2b. Light purple or brownish purple. (See *3a*, sec. 2, p. 87.)
  ***T. dubium*, suckling clover (pl. XXXIX, 342).

1b. Seeds dull or lustrous.

3a. Light to dark purple, at least in part, occasional seeds dark olive-green in *michelianum* (*5a*).

4a. Seeds triangular or ovate, the radicle much shorter than the cotyledons, the hilum in a shallow indentation on the side; color yellow on the narrower end and shading to purple below, sometimes entirely purple. (See *13a*, sec. 2, p. 88.)

**T. pratense*, red clover (pl. XXXIX, 347).

4b. Seeds oval, commonly with a short point-like projection at the base of the cotyledons in *lappaceum* (*5b*); length 1½–1¾ mm., width 1½ mm. or more.

5a. Radicle equal to cotyledons or nearly so. (See *14a*, sec. 2, p. 88.)

***T. michelianum*, big-flowered clover (pl. XXXIX, 345).

5b. Radicle markedly shorter than the cotyledons, the hilum in a shallow indentation on the side; surface minutely uneven or lumpy, commonly with faint longitudinal wrinkles on the tips of the cotyledons; occasional seeds entirely yellow. (Compare with *5b*, sec. 2, p. 87.)

***T. lappaceum*, lappa clover (pl. XXXIX, 344).

3b. Black or nearly so; orbicular or broadly oval and rotund; hilum in a distinct notch; length and width about equal, 2½–3 mm.

**T. subterraneum*, subterraneum clover (pl. XL, 363).

TRIGONELLA

Species of *Trigonella* vary greatly in size, shape and surface character. The genus is represented here by two species. One species, fenugreek, is in limited cultivation in the United States. Other species are sometimes found incidentally with crop seeds of European or Asiatic origin.

Trigonella foenum-graecum, fenugreek (pl. XL, 364).—Seeds are oblong and angular, with a deep groove between the radicle and cotyledons, the length is 5½–6 mm., width 3–3½ mm. Color is light grayish brown and the surface is faintly roughened by minute tubercles or short raised lines. The radicle is about half the length of the cotyledons and the minute hilum lies partly obscured within a deep notch. Seeds have the characteristic odor of fenugreek.

Trigonella polycerata, (no common name).—Seeds are narrowly oblong, 2–2½ mm. long, 1 mm. or more wide. Surface is roughened by minute blister-like elevations. The radicle is slightly shorter than the cotyledons, with only a faint line between the radicle and cotyledons. The minute hilum lies in a shallow indentation near one end of the seed. This species may occur with such crops as alfalfa or black medic.

Trigonella ornithopodioides (*Falcatula ornithopodioides*) has been referred to *Trifolium ornithopodioides* and is described under that genus. This seed may occur in white clover of New Zealand origin.

VICIA

Refer to fig. 12 p. 66 for structural features of legume seeds.

Eighteen species of vetch (*Vicia*) are of interest in our agriculture. Eight species are cultivated crops. Some of the species have several horticultural varieties which are usually not distinguishable from the species. The crop species are indicated by an asterisk (*) in the seed key that follows. Ten species are classed as weeds. These may occur as contaminants in vetch and similar crops. Three of the weed species, *V. hybrida*, *melanops*, and *lutea* are sometimes found in imported seed.

The position of the chalaza is an important diagnostic feature in seeds of vetch. It appears on the seed surface as a dark-colored, slightly raised spot. (See pl. XLI, 374.) The hilum is another important feature. An examination of the illustrations of the seeds will show that most species have a characteristic type of hilum.

Similar-Appearing Species

V. sativa, common vetch, and *V. angustifolia,* narrowleaf vetch.—There are several horticultural varieties of common vetch and the seeds may vary considerably in size and color. The seed key describes the type that is common in commercial channels. Some seeds of common vetch are very similar to narrowleaf vetch and the two cannot always be distinguished.

Narrowleaf vetch is represented by two types of seed. One type is black, the other is greenish with dark mottling. Both types may occur in domestic seed. The green, mottled seed has not been observed in imported seed.

V. villosa, hairy vetch, and *V. dasycarpa,* woollypod vetch.—Seeds of the two species are somewhat similar and may be confused. A careful examination of the size and shape of the hilum will show that the two species are readily distinguishable.

Key to *Vicia*

1a. Chalaza on the back, opposite the hilum.
2a. Hilum about 2 mm. long, flush with the seed surface or nearly so.
3a. Hilum ½–¾ mm. wide, narrowly oblong; seed 4–4½ mm. long.
**V. pannonica,* Hungarian vetch (pl. XL, 367).
3b. Hilum about 1¼ mm. wide, short-oval; seed 5 mm. long.
V. hybrida, no common name (pl. XLI, 370).
2b. Hilum about 4 mm. long, oblong.
4a. Hilum narrow, ½–¾ mm. wide, slightly depressed at the margins; seed 6–6½ mm. long............. *V. lutea,* yellow vetch (pl. XLI, 369).
4b. Hilum wide, 1 mm. or more, flush with the seed surface or nearly so; seed 4½–5 mm. long........... *V. melanops,* black vetch (pl. XL, 368).
1b. Chalaza near one end of the hilum.
5a. Hilum linear, obscured by a persisting whitish tissue sometimes lacking in *7a;* seeds 3–4 mm. long.
6a. Hilum about ⅔ of the circumference of the seed; seeds lens-shaped, brownish and faintly mottled.
V. grandiflora, showy vetch (pl. XLI, 373).
6b. Hilum much shorter, 1–3 mm. long; seeds slightly flattened but not lens-shaped.
7a. Hilum 2½–3 mm. long; seeds dull black; seeds from occasional plants lack the whitish appendage on the hilum.
**V. atropurpurea,* purple vetch (pl. XLI, 371).
(*V. bengalensis.*)
7b. Hilum about 1 mm. long; seeds brownish, may be faintly mottled.
**V. articulata,* monantha vetch.
5b. Hilum not obscured, or only partially so in *hirsuta.*
8a. Seeds relatively small, length 1½–2 or 3 mm.
9a. Seeds roughened by knob-like protuberances and minutely stippled; seed flattened at each end of the hilum, 4-sided in outline, length 2 mm. Hilum short-oval, length 1 mm., width ½ mm., dark reddish brown; seed light grayish brown, with a prominent chalaza near the hilum.
V. lathroides, lowpea vetch.
9b. Seeds smooth, mostly spherical.
10a. Hilum short-oblong, length 1¼ mm., width ½ mm. or less; seed reddish brown or greenish and finely mottled, length 2–2½ mm.
V. tetrasperma, fourseed vetch (pl. XL, 366).

10b. Hilum markedly longer, 2–3 or 3½ mm.
11a. Hilum linear, width ½ mm. or less.
12a. Seeds glossy, yellowish or brownish and copiously mottled with purple; hilum 2–2½ mm. long, commonly with a stalk-like appendage loosely persisting on the hilum.
V. hirsuta, tiny vetch (pl. XL, 365).
12b. Seeds dull, black or lighter and mottled; hilum 2½–3 mm. long ---------------------- *V. cracca*, cow vetch (pl. XLI, 372).
11b. Hilum linear-ovate, width at the broader end about ¾ mm.
13a. Hilum about 2 mm. long, slightly depressed at the margins and raised along the median groove; seeds black and lustrous or dull greenish with dark mottling.
**V. angustifolia*, narrowleaf vetch (pl. XLI, 375).
13b. Hilum about 3½ mm. long, the ends wider than at the middle, flat, grayish and scurfy; seeds dark brown with obscure mottling ---------------------- *V. americana*, American vetch.
8b. Seeds relatively large, length 4–5 or 6 mm., up to 14 mm. in *faba*.
14a. Chalaza 2½–3 mm. from end of hilum; hilum about 1 mm. long, ½ mm. wide.
Seeds dull, pale brownish and usually mottled with dark green; the side opposite the chalaza is flattened at right angles to the hilum so that seed appears triangular in outline.
V. ervilia, bitter vetch.
14b. Chalaza 1 mm. or less from end of hilum.
15a. Seeds spherical or slightly flattened.
16a. Hilum oval, flush with the seed surface, smooth and flat, black or reddish, length about 2 mm., width 1 mm.; seed black or obscurely mottled ------- **V. villosa*, hairy vetch (pl. XLI, 377).
**V. villosa* var. *glabrescens*, smooth vetch.
16b. Hilum depressed at the margins.
17a. Hilum narrowly oval, flat, the surface slightly granular and scurfy with a light-colored strip along the median groove, length about 2 mm., width ¾ mm. or less; seed color dull black or obscurely mottled.
**V. dasycarpa*, woollypod vetch (including the vars. Lana and Auburn) (pl. XLI, 378).
17b. Hilum linear-ovate, raised along the median groove, length 2–2½ mm., width ½ mm.; seed color variable, commonly reddish brown and obscurely mottled. A variable species; may intergrade with *angustifolia*, *13a*.
**V. sativa*, common vetch (pl. XLI, 374).
15b. Seeds oblong or broadly oval, thick, length 13–14 mm., width 8–9 mm.; seed color reddish brown; hilum 5–6 mm. long, 1½ mm. wide, lies at one end on the short axis of the seed.
**V. faba*, broadbean.

Vigna

Vigna sinensis, cowpea.—There are many varieties of cowpea in cultivation. The seeds vary widely in color and size, and to some extent in shape. The hilum is about 2–3 mm. long, the width and shape varying somewhat with the variety. The entire hilum is depressed and obscured by a whitish corky tissue which is elevated above the surface of the seed. The outer portion of this whitish elevation is darker in color, usually greenish.

The genus is represented here by two varieties: Brabham (pl. XLI, 380), showing a smooth, mottled seed, and Blackeye (pl. XLI, 379), showing a wrinkled seed surface and a characteristic black band around the hilum.

Vigna and *Phaseolus* compared.—Seeds of these two genera are somewhat similar. The following observations may serve as a guide in distinguishing them.

Vigna: The hilum is broadly ovate or oval, short and broad in proportion to its length; the greenish band surrounding the elevated whitish layer covering the hilum is wide and prominent.

Phaseolus: The hilum is narrowly oblong or oval, narrow in proportion to its length; the whitish layer covering the hilum is flatter than in *Vigna* and the greenish tissue around it is narrow and inconspicuous or lacking.

CRUCIFERAE—MUSTARD FAMILY

Fruits in the mustard family are usually 2-celled pods, the two valves separating at maturity. The pods may be long and narrow as in winter rape, or they may be short and broad as in some of the mustards. A few species have pods that are indehiscent and nutlike as in giant mustard (*Rapistrum rugosum*), or they may separate crosswise into 1-seeded segments as in wild radish (*Raphanus raphanistrum*) pl. XXXII, 252. Thus the seed unit may be a true seed, an indehiscent pod, or a segment of a pod.

A large embryo fills the entire seed cavity and there is no visible endosperm. The embryo is curved or folded in various ways and its position is often helpful in making an identification. Figure 16 shows three types of embryo as they may appear in longi-section and in cross section.

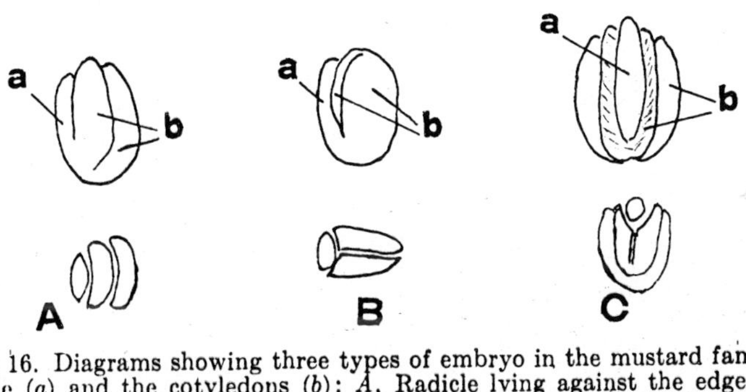

FIGURE 16. Diagrams showing three types of embryo in the mustard family, the radicle (*a*) and the cotyledons (*b*): *A*, Radicle lying against the edges of the cotyledons; *B*, radicle lying against the flat side of the cotyledons; and *C*, cotyledons folded and partially enclosing the radicle.

The outline of the radicle and cotyledons is often plainly visible on the seed surface, and the type of embryo can be determined in many cases without sectioning the seed. Many of the illustrations show the embryo type that is characteristic of a species, and this feature will not be included in the seed keys and descriptions.

Many species of the mustard family are of agricultural importance both as vegetable and field crops and field weeds. In addition to the crop species, only the more important weed species are described here.

BRASSICA

The genus *Brassica* includes most of the agricultural and vegetable crops in the mustard family, as well as several noxious weeds.

Winter rape (*B. napus* var. *biennis*) is an important fodder crop. Most of the seed is imported. The oil-seed rapes, summer rape (*B.*

600705 O—62——7

napus var. *annua*) and the turnip-rapes (*B. campestris* vars.), are not grown in this country, but the seed is imported for processing. Some of the seed of the vegetable species (mostly in the *B. oleracea* group) and the species used in the manufacture of condiments are grown here and some are imported. The latter includes such species as *B. nigra, hirta*, and *juncea*. Field kales are a very minor crop in this country. The variety Thousand-Headed kale is sometimes grown locally as a fodder crop. Siberian kale is planted for table use. Charlock (*B. kaber*) is classed as a noxious weed in about half of the States.

Brassica Classification

The species of *Brassica* fall into three distinct groups: (1) The rapes' (2) cabbages or coles, and (3) the mustards. Based on seed character' the species in group 3 may be further divided into three subgroups· There is a striking similarity in the seeds within each of the groups and subgroups, and individual seeds are not always distinguishable with certainty. In such cases a greenhouse test should be made to verify the seed identification. Under favorable conditions, identification of the seedlings is usually possible in about 14 days.[2,3]

Confusion in nomenclature of the species is often encountered. The following outline may serve as a guide to the classification used in this publication.

Group 1, The Rapes

a. Winter rape, *B. napus* var. *biennis* (Schubl. & Mart) Reichb.
b. Summer rape, *B. napus* var. *annua* Koch.
c. Rutabaga, *B. napus* var. *napobrassica* (L.) Reichb.
d. Rape-like kales, Siberian kale, *B. napus* var. *pabularia* (DC.) Reichb.

Group 2, Cabbages

Cabbage and allies, including the field variety Thousand-headed kale, *B. oleracea* L. and varieties.

Group 3, The Mustards

a. Turnip and allies:
- Wild turnip, *B. campestris* L.
- Mediterranean wild turnip, *B. tournefortii* Gouan.
- Turnip-rape, annual, *B. campestris* L.
- Turnip-rape, biennial, *B. campestris* var. *autumnalis* DC.
- Turnip, edible, *B. campestris* var. *rapa* (L.) Hart. or *B. rapa* L.
- Table mustards:
  - Spinach mustard, *B. perviridis* Bailey.
  - Chinese cabbage, *B. pekinensis* (Lour.) Rupr.
  - Pakchoi, *B. chinensis* L.

b. Black mustard, *B. nigra* (L.) Koch.
Brown mustard and Oriental mustard, *B. juncea* (L.) Cosson.
White mustard, *B. hirta* Moench, or *B. alba* Moench.
c. Charlock, *B. kaber* (DC.) Wheeler, or *B. arvensis* (L.) Rabenh.

Similar-Appearing Species

Group 1, the rapes.—Individual seeds, and often the bulk samples, of summer rape and the rape-kales may be indistinguishable from winter rape. Since summer rape is an oil-seed variety, it is usually important to distinguish it from the forage-plant variety, winter rape. It may be

[2] Musil, A. F. Distinguishing the Species of *Brassica* by their Seed. U.S. Dept. Agr., Misc. Pub. No. 643. 35 pp., illus. 1948.

[3] Musil, A. F. Identification of Brassicas by Seedling Growth or Later Vegetative Stages. U.S. Dept. Agr., Circ. No. 857. 26 pp., illus. 1950.

necessary to make a greenhouse test for positive identification.[3] Rutabaga seeds are usually distinguishable both individually and in the bulk.

Group 2, cabbages.—There are many botanical and horticultural varieties of *B. oleracea*. These include such plants as cabbage, collards, cauliflower, broccoli, Brussels sprouts, kohlrabi, and the curled kitchen kales. The structural features of the seeds are so similar that it is usually not possible to distinguish them with certainty. A greenhouse test may be necessary to determine the variety.

Group 3, turnip and allies (subgroup a).—The seeds are distinctive as a group, but the seeds of annual and biennial turnip-rape and the garden turnips are usually not distinguishable. There are many horticultural varieties of garden turnip, and variations in seed character may be expected. In some areas garden turnip escapes from cultivation and the seeds occur as weed contaminants with crop seeds.

Seeds of the three table mustards, spinach mustard, chinese cabbage, and pak-choi, show a relationship to the *campestris* group. With practice is is usually possible to distinguish the bulk samples. There are several horticultural varieties of chinese cabbage and these do not appear to be distinguishable by seed character. Because these three species are in rather limited use as vegetable crops, descriptions of the seeds are not included here.

Mustards (subgroup b).—This group includes *B. nigra, B. juncea*, and *B. hirta*. The three species are readily distinguished.

Charlock (subgroup c).—As will be shown in the seed key, charlock seed has structural features that are not found in any of the other species. It is possible to identify this seed even when it is poorly developed.

Structural Features of the Seeds

The seeds of *Brassica* are mostly spherical, or sometimes slightly flattened. The surface has a reticulum or netting of lines or ridges, and the character of these ridges is an important diagnostic feature. The reticulum should be viewed along the median portion of the seed. The areas between the lines of the reticulum are referred to in the key as "interspaces." It is not practical to assign measurements to these microscopic structures, and the terminology used in describing them is of necessity relative. With experience, these relationships become evident.

In addition to the reticulum, the seed surface is covered with microscopic pits. These pits are usually covered with a whitish film, giving the appearance of white spots on the seed surface. The term "stipples" is used in describing them.

The length of seed may be determined by a line perpendicular to the hilum. In the case of spherical seeds, the length and width are approximately equal. As in other kinds, seeds of *Brassica* may vary in size within a species, but with experience a seed can usually be placed in either a large-seeded or a small-seeded group. When in doubt as to the correct size, a check of other features will quickly indicate the proper category in the seed key.

The descriptions in the seed key are based on a magnification of 40 ×. Seeds of charlock are easily recognized by their smooth and velvety appearance under the low magnification of a 7 × hand lens.

Poorly developed seeds, however, would require the higher magnification to evaluate the minute structural features.

Key to *Brassica*

1a. Seeds yellow or pale straw-color.
2a. Stipples distinct, shiny; seeds small (less than 2 mm. long).
3a. Reticulum of fine, distinct lines, the interspaces large and shallow.
B. juncea, Oriental mustard (pl. XXXV, 267).
3b. Reticulum lacking or only faintly evident.
B. campestris var. *Sarson*, Sarson.
2b. Stipples not evident; a thick reticulum with very small interspaces obscured by a translucent mucilaginous layer; seeds 2–3 mm. long.
B. hirta (*B. alba*), white mustard (pl. XXXIV, 263).
1b. Seeds grayish black, dull brown or reddish, never yellow.
4a. Seeds large (2–3 mm. long).
5a. Stipples evident on the ridges of the reticulum.
6a. Interspaces of reticulum large.
7a. Lines of reticulum flattish, seed appears rather smooth; stipples small, round or rectangular, shiny or dull.
B. napus var. *biennis*, winter rape (pl. XXXIII, 258).
Reticulum coarser, stipples larger and more prominent than above.
B. napus var. *annua*, summer rape (pl. XXXIII, 259).
B. napus var. *pabularia*, Siberian kale (pl. XXXIII, 260).
7b. Lines of reticulum are coarse ridges; stipples large, rectangular and shiny, prominent on the broad ridges.
B. napus var. *napobrassica*, rutabaga (pl. XXXIII, 261).
6b. Interspaces of reticulum very small, the reticulum flattish and obscure; seeds compressed laterally, appear quite smooth; color reddish, apparently never black.
B. campestris var. *rapa*. (*B. rapa*), Shogoin turnip (pl. XXXIV, 264).
5b. Stipples not evident on the reticulum; lines of reticulum narrow, often semi-glossy; interspaces very small, the stipples minute and partially obscured; seeds tend to be flattened or warped; color bronze or grayish black................*B. oleracea*, cabbage and allies (pl. XXXIV, 262).
4b. Seeds small (less than 2 mm. long).
8a. Stipples evident on the coarse ridges of the reticulum; interspaces very small, or only rarely large; stipples small, roundish, shiny and prominent.
B. campestris (including *B. rapa*), turnip and turnip-rapes (pl. XXXIV, 265).
8b. Stipples not evident on the reticulum.
9a. Reticulum fine, indistinct, may appear to be lacking; interspaces very small; stipples very minute and partially or completely obscured; color black............*B. kaber* (*B. arvensis*), charlock (pl. XXXV, 268).
9b. Reticulum prominent.
10a. Interspaces very small; lines of reticulum whitish; stipples minute; seeds light reddish brown or black, spherical, very small, 1–1½ mm. in diameter..................................*B. tournefortii*, Mediterranean wild turnip (pl. XXXV, 269).
10b. Interspaces large; color light to dark reddish brown.
11a. Reticulum of thick, prominent ridges; interspaces deep, concave, glossy; stipples minute and partially or completely obscured.
B. nigra, black mustard (pl. XXXV, 266).
11b. Reticulum of fine, distinct lines; interspaces shallow, not concave; stipples not obscured, small but larger than in *nigra*, dull or shiny.
B. juncea, brown or India mustard (pl. XXXV, 267).

BARBAREA

Two species of *Barbarea* are commonly found in agricultural seed, *B. verna* and *B. vulgaris*. The latter species is classed as a noxious weed in several Eastern States.

One side of the seed is usually rounded, the other side is humped up across the middle and slopes down to each end. The oblong seeds are dark brown, overlaid by a yellowish, waxy and translucent film. The size of the seeds and the height of the reticulum are the main distinguishing features of the two species. The differences in surface configuration are usually evident at a magnification of 6 × or 10 ×.

Barbarea verna, early wintercress (pl. XXXII, 243).—Seeds are 2–2½ mm. long, 1–1½ mm. wide. The surface has a high reticulum of thin dark brown lines, and the interspaces are narrow, deep and glossy. The waxy, yellowish film produces a honeycomb effect.

Barbarea vulgaris, bitter wintercress or yellow rocket (pl. XXXII, 244).—Seeds are 1½ mm. long, 1 mm. wide. The reticulum is not prominent, sometimes obscured, and the interspaces are shallow. The yellowish film produces a bubbly surface.

CARDARIA

Two species of *Cardaria* are classed as noxious weeds: *Cardaria draba* and its var. *repens* (formerly *Lepidium draba*), and *C. pubescens* (formerly *Hymenophysa pubescens*).

Pods of the three kinds are readily distinguished (see pl. XXXII, 246–247). Well-developed, mature seeds of the two species are usually distinguishable, but seeds found with crop seeds are not always fully developed and such seeds may be indistinguishable.

The bulk aspect of seeds of *C. draba* tend to be a dark reddish brown, those of its var. *repens* and *C. pubsecens* tend to be lighter in color. The degree of redness, however, may vary with conditions of development. In all three kinds the radicle of the embryo lies against the edges of the cotyledons (see fig. 16, *a*, p. 93). The length is approximately the same, 2½–3 mm. The shape in outline is similar, but *pubescens* tends to be more broadly oval, flatter and smoother than the other two. This difference in surface is usually evident at a magnification of 6 × or 10 ×.

The following features appear to be fairly characteristic, but in all cases allowance should be made for variations in size, shape, and color due to stage of maturity and conditions of development.

Cardaria draba, hoary cress, including the var. *repens,* lens peppercress. (pl. XXXII, 246).—Seeds are oval, about 2 mm. wide, the cotyledon edge thicker than the radicle edge. The surface of the seed is roughened by minute reticulations.

Cardaria pubescens, ballcress (pl. XXXII, 247).—Seeds are broadly oval, about 2¼ mm. wide, the thickness about uniform throughout. The reticulum on the seed surface is so minute the seed appears to be quite smooth.

CHORISPORA

Chorispora tenella, blue mustard (pl. XXXII, 248).—The long pods of this species break up at maturity into indehiscent segments. If the seed were removed from the woody segment it would reveal the features that are characteristic of the mustard family. The rather mealy seed is not likely to be found in crop seeds.

The seed unit is a segment of the pod. The segments are rectangular in shape, 2–2½ mm. long and 1¾–2 mm. wide. One side is flat and smooth; the other side is arched, with a wrinkled, corky strip between wide, smooth margins. These segments may occur in crop seeds from States west of the Mississippi River. The species is classed as a noxious weed in two Western States.

LEPIDIUM, CONRINGIA, CAPSELLA

There are several species of *Lepidium* occurring as field weeds. Two species are of special interest. Field cress (*L. campestre*) is a common impurity in various crop seeds. Perennial mustard (*L. latifolium*) is locally established in the Pacific Northwest where it is classed as a noxious weed. In the seed key that follows, these two species are compared with the similar-appearing seeds of *Conringia orientalis*, hares-ear-mustard, and *Capsella bursa-pastoris*, shepherds-purse.

It will be observed in the illustrations that the radicle of the embryo lies against the flat side of the cotyledons in all these species. The color of the seeds is light reddish brown or a dull dark brown. The difference in surface configuration can usually be recognized at a magnification of 6 × or 10 ×.

Key to *Lepidium, Conringia, Capsella*

1a. Length about 1¼ mm., or less.
- *2a.* Seeds oblong, thin and flat, the thickness uniform throughout, the ends bluntly rounded; reticulations mostly fine lines, the interspaces elongated longitudinally.................. *Capsella bursa-pastoris*, shepherds-purse.
- *2b.* Seeds oval or ovate, the thickness not uniform, thicker on the end opposite the hilum; reticulations faint, seed appears smooth under a hand lens. *Lepidium latifolium*, perennial mustard (pl. XXXII, 251).

1b. Length 2–3 mm., width 1 mm. or more; cotyledons wide and flattened on the back so that seed stands firmly on a plane surface.
- *3a.* Seed surface relatively smooth, appears bubbly under magnification; radicle thick, with a slight groove on either side, the tip usually lying off-center.......... *Conringia orientalis*, hares-ear-mustard (pl. XXXII, 249).
- *3b.* Surface roughened by minute tubercles; radicle flattened out toward the tip, does not lie off-center. *Lepidium campestre*, field cress (pl. XXXII, 250).

RAPHANUS

Two species of *Raphanus* are found quite commonly with crop seeds. *Raphanus raphanistrum*, wild radish (pl. XXXII, 252), is classed as a noxious weed in most Eastern and Southern States. *Raphanus sativus*, radish (pl. XXXII, 253), escapes from cultivation and may occur as a contaminant in grain fields.

Seeds of both species are light to dark brown or yellowish brown, oval or ovate, thick and slightly flattened laterally. The hilum is a rather indefinite dark spot at the broader end of the seed. The surface has a reticulum of fine lines, the interspaces are shallow, often partly glazed over and tend to be elongate. Stipple points are dull and minute and scarcely evident at a magnification of 40 ×. Size of the seeds is about the same for both species, length 3–4½ mm., width 2½–3 mm. The seeds of wild radish usually average somewhat smaller than those of the cultivated radish. Without the pods, individual seeds appear to be indistinguishable.

Pods of *Raphanus raphanistrum*, wild radish, are linear, strongly ribbed and constricted between the seeds like a string of beads. The pods break up at maturity on the constrictions between the seeds into cylindrical segments. These bony, indehiscent segments are the units commonly present in crop seed samples.

Pods of *Raphanus sativus*, radish, are broader than wild radish, longitudinally nerved but not ribbed, and obscurely or not at all constricted between the seeds. The pods are spongy in texture and although indehiscent, they break open easily in processing and the seed unit is the hulled seed.

RORIPPA

One species, *Rorippa austriaca*, yellow fieldcress, is considered a noxious weed in several Western States. It is reported to be a light seeder under some conditions and the seed would not be likely to occur frequently with crop seeds. Its seeds are compared here with the similar-appearing seeds of *R. sylvestris*, creeping yellowcress, which is rarely found in crop seeds.

Seeds of both species are very small and without the pods they may be difficult to distinguish. Both are light reddish brown in color and broadly oval in shape. The interspaces of the reticulum are shiny. The surface configuration is usually distinguishable under a magnification of 40 ×.

R. austriaca, yellow fieldcress (pl. XXXII, 254).—Seeds are produced in a round pod and they are often variously bent or misshapen. The length and width are about equal, ¾–1 mm. The surface has a distinct, coarse reticulum.

R. sylvestris, creeping yellowcress (pl. XXXII, 255).—Seeds are produced in a long, straight pod and are symmetrically oval in shape. The length and width are about equal, ½–¾ mm. The reticulum is very minute and rather obscure.

SISYMBRIUM

Seeds of *Sisymbrium* are not considered noxious weeds, but they may be objectionable when occurring in quantity. The following tabulation describes four weed species that occur commonly with field seeds: *S. orientale*, oriental hedgemustard; *S. altissimum*, tumblemustard; *S. officinale*, hedgemustard; *S. irio*, London rocket.

Four Species of *Sisymbrium* Compared

Approximate Size

S. orientale, oriental hedgemustard.—Length 1 mm., width ¾ mm.

S. altissimum, tumblemustard (pl. XXXII, 256).—Length 1 mm., width ½ mm.

S. officinale, hedgemustard (pl. XXXII, 257).—Length 1 mm., width ½ mm. or larger.

S. irio, London rocket.—Length 1¼ mm., width ½ mm.

Shape

S. orientale, oriental hedgemustard.—Oblong, thick and blocky, one sidesmoothly rounded, the other side humped up about midway and sloping to each end; the end opposite the scar truncate or obtusely rounded; a prominent groove between radicle and cotyledons.

S. altissimum, tumblemustard.—As in *orientale* but flatter.

S. officinale, hedgemustard.—As in *orientale* but flatter; a fine groove between radicle and cotyledons.

S. irio, London rocket.—Narrowly oblong and scarcely flattened, the sides smoothly rounded.

Color and Surface

S. orientale, oriental hedgemustard.—Almost smooth, waxy; the cotyledon area dark bluish green, the radicle area yellowish.

S. altissimum, tumblemustard.—Almost smooth, waxy; light yellowish or greenish, with two prominent green or brownish lines on each side, outlining the edges of the cotyledons.

S. officinale, hedgemustard.—Faintly striate, dull or fresh seed waxy; reddish brown or dark greenish brown, the two greenish or brownish lines on the sides lacking.

S. irio, London rocket.—Smooth and glossy; yellow or light buff.

FAMILIES CONTAINING NOXIOUS-WEED SPECIES

Since this publication is concerned primarily with the identification of crop seeds, it is not practical to include all species classed as noxious weeds under State seed laws. Ten families have been selected which contain the more important noxious weeds or those that may be difficult to distinguish from the relatively harmless species.

CYPERACEAE—SEDGE FAMILY

The sedge family is represented in our agriculture by several genera of field weeds, such as *Cyperus*, *Carex*, *Scirpus*, and *Eleocharis*. Except for *Cyperus*, the weeds are of relatively minor importance. *Cyperus rotundus*, nutgrass, and *C. esculentus*, yellow nutgrass, are considered noxious weeds in many States, particularly in the South and Southwest.

The seed unit in the sedges is a 1-seeded, indehiscent achene. The endosperm fills almost the entire seed cavity. The embryo is minute and is located at the base of the seed. The achene of *Carex* is produced within a beaked, sac-like structure which sometimes persists in processed seed. (See pl. XXX, 206.) Achenes of *Cyperus* usually fall free from their enveloping bracts.

Nine species of *Cyperus* that are known to have been found in field seeds are described in the seed key.

Key to *Cyperus*

1a. Achenes flattened, broadly obovate (urn-shaped); length 1 mm. or less, width ¾ mm. or less.
 2a. Dark brown, overlaid with a fine, grayish reticulum. *C. diandrus*, cypressgrass.
 2b. Black, the surface with fine vertical lines and whitish, broken, transverse ridges ---------- *C. flavescens*, yellow cyperus.
1b. Achenes 3-sided.
 3a. Length 1–1¼ mm., width ¾–1 mm.
  4a. Narrowly obovate, with a prominent knob at the base, the surface finely striate; color grayish brown ---------- *C. eragrostis*, umbrella sedge.
  4b. Broadly obovate, the three sides of achene equal and smooth. *C. compressus*, flat sedge.
 3b. Length 1½–2 mm., width varies with the species.
  5a. Surface tubercled; length 2 mm.
   6a. Width less than ½ mm.; linear oblong, dark reddish brown; tubercles in distinct vertical lines ---------- *C. strigosus*, straw-colored cyperus.
   6b. Width 1 mm. or less.
    7a. Broadly elliptic, the ends obtuse; dark brown, dull; tubercles close-spaced ---------- *C. filiculmis*, slender cyperus.
    7b. Narrowly oblong, the ends pointed; light reddish brown; tubercles more wide-spaced than in *filiculmis* ----- *C. ovularis*, globose cyperus.

  5b. Surface other than tubercled.
   8a. Surface with a fine, distinct reticulum; grayish brown, dull; two sides of achene equal, the third side broader; length 1½–2 mm., width 1½ mm ---------- *C. rotundus*, nutgrass (pl. XXX, 208).
   8b. Surface bubbly, amber-colored or light brown and lustrous; the three sides of achene equal; length 1½ mm., width ½–¾ mm. *C. esculentus*, yellow nutgrass (pl. XXX, 207).

LILIACEAE—LILY FAMILY

Most of the weed species of the lily family that invade cultivated fields are relatively unimportant. The most troublesome by far are the two species of wild onion: *Allium vineale*, wild garlic, and *Allium canadense*, wild onion. Both species are usually classed as noxious weeds.

The seed unit in wild onion and wild garlic is a bulblet. The unit consists of a solid bulblet within a loose-fitting outer husk that is whitish or yellowish and lustrous. The growing point is located at the base of the solid bulblet. Bulblets found in processed crop seeds may be partly or entirely devoid of the outer husk, and the solid bulblet may be shriveled or variously damaged. In this condition it may be difficult to determine the species. Well-developed and undamaged bulblets may be identified by shape and size. Wild garlic and wild onion also produce underground bulbs, rarely true seeds; but these bulbs and seeds are not known to occur in crop seeds.

Allium vineale, wild garlic (pl. XXX, 209).—Bulblets are obovate or top-shaped, terminating in a slender beak and tapering to a narrow base. The growing point is at the narrow basal end. The size of bulblets may vary considerably, but the better developed bulblets may be 4–7 mm. long. The solid inner bulblet is normally obovate but due to stage of maturity at time of harvest, it may be more or less shriveled or misshapen.

Allium canadense, wild onion.—The shape of the bulblets is ovate, broadest below the middle and tapering to a pointed apex. The growing point is at the broad basal end of the bulblet. Size of bulblets is variable but, in general, they are markedly larger than *vineale*. The length may be up to 10 or 15 mm.

POLYGONACEAE—BUCKWHEAT FAMILY

The buckwheat family is represented here by two genera, *Rumex* and *Polygonum*. Both genera contain noxious-weed species, the seeds of which are quite similar to some of the harmless species.

The fruit or so-called seed is an achene. The hard outer hull (pericarp) is brown and glossy in most species and the calyx, or some portion of it, sometimes persists on the mature fruit. The seed coat is brown and tissue-like. The embryo lies in a groove on the outside of an abundant whitish endosperm. The pericarp and seed coat of *Rumex* often become removed in processing, and only the trigonous endosperm and embryo remain with the crop seed.

The shapes of the achenes and the position of the embryo are shown diagrammatically in figure 17. The position of the embryo is an important feature distinguishing the two genera.

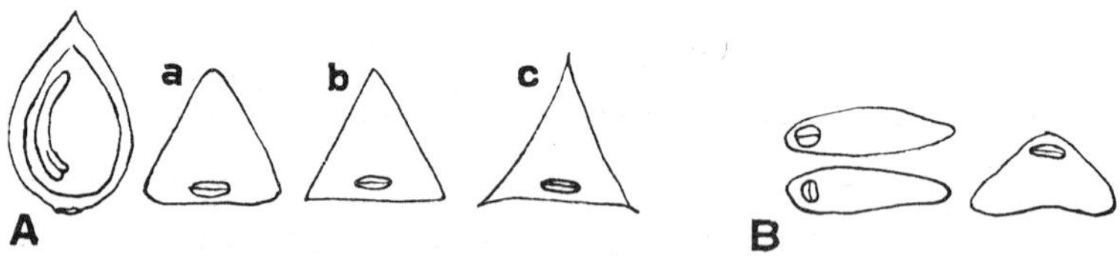

FIGURE 17.—(*A*), Achenes of *Rumex* in longi-section and cross section, showing position of embryo and three types of angles of the trigonous achenes: (*a*), obtuse; (*b*), acute; (*c*), minutely winged. (*B*), Achenes of *Polygonum* in cross-section, showing two shapes of achenes and the position of the embryo.

Distinguishing the Genera *Polygonum* and *Rumex*

Rumex (fig. 17).—Achenes are 3-angled, the three sides equal or, more rarely, the achene is slightly flattened. The surface is smooth and glossy. The embryo lies in a groove or channel along the middle of one of the sides of the endosperm.

Polygonum (fig. 17).—Achenes are flattened, or if 3-angled, two of the sides are equal and the third side is broader. The surface may be smooth and glossy or roughened and dull. Regardless of the shape, the embryo lies in a groove or channel on one of the angles of the endosperm.

Similar-Appearing Species

Rumex species.—The small-seeded species of *Rumex* are easily recognized. Species having achenes that are 2–3 mm. long may be very similar and difficult to distinguish with certainty. This is especially true of such species as *crispus*, *obtusifolius*, and *pulcher*. The seed key describes the more characteristic seed types of each species but intergrading forms are found on most plants. These intergrading forms are usually indistinguishable.

Polygonum species.—*Polygonum argyrocoleon*, silversheath knotweed, because of its 3-sided shape can be mistaken for a *Rumex*. A careful comparison of the generic features as described in the seed key will readily distinguish it. When in doubt, a cross section of the achene will indicate the genus by the position of the embryo.

Polygonum aviculare, knotweed, frequently produces cleistogamous fruits (pl. XXX, 212). These achenes are sometimes found in crop seeds. They are smooth and glossy and may superficially resemble *Rumex*. It is sometimes necessary to section doubtful achenes to determine the genus.

The measurements given in the seed keys for both *Rumex* and *Polygonum* should be considered as approximate only. The measurements represent the usual size of seeds, but it can be expected that seeds will be found that are somewhat larger or smaller than those stated. The widths represent the widest point of the seed.

Key to *Rumex*

1a. Achenes small, length 1½ mm. or less, width 1 mm. or less.
- *2a.* Narrowly elliptic, both ends equally long-pointed, the angles acute; color pale brown; length 1¼–1½ mm., width ½ mm.
  R. persicarioides, golden dock (pl. XXXI, 225).
- *2b.* Broadly elliptic.
  - *3a.* Ends of achenes blunt and scarcely pointed, the angles obtuse; dark reddish brown; the calyx usually persisting; length 1½ mm., width 1¼ mm.
    R. acetosella, sheep sorrel (pl. XXXI, 219).
  - *3b.* Ends of achenes equally short-pointed, the angles acute; yellowish brown and translucent, with a distinct dark brown line on each angle; length 1½ mm., width 1 mm. or less................*R. hastatulus*, sourgrass.

1b. Achenes markedly larger.
- *4a.* Angles of achenes obtuse or if acute, not distinctly winged.
  - *5a.* Base of achenes minutely stipitate.
    - *6a.* Color reddish brown, glossy, darker brown on the angles; achenes taper abruptly to a short-pointed apex; length 2–2½ mm., width 1½–1¾ mm................*R. crispus*, curly dock (pl. XXXI, 222).
    - *6b.* Color grayish brown or chocolate brown, lustrous, not darker on the angles; achenes taper rather evenly to a short-pointed apex; length 2½ mm., width 1¼ mm.
      R. obtusifolius, broadleaf or bitter dock (pl. XXXI, 223).
  - *5b.* Base of achenes broad, not stipitate; shape triangular; color dark reddish brown, glossy.
    - *7a.* Achenes about 1¾ mm. long, 1½ mm. wide.
      R. conglomeratus, clustered or green dock (pl. XXXI, 221).
    - *7b.* Achenes about 2½ mm. long, 1½–1¾ mm. wide.
      R. pulcher, fiddleleaf dock (pl. XXXI, 226).
- *4b.* Angles of achenes acute, appear narrowly winged.
  - *8a.* Wings of angles prominent.
    - *9a.* Achenes elliptic, the apex and base equally pointed; length 3½–4 mm., width 2¼ mm.....*R. occidentalis*, western dock (pl. XXXI, 224).
    - *9b.* Achenes broadly ovate, the apex short-pointed.
      - *10a.* Achenes about 6 mm. long, 3½–4 mm. wide.
        R. venosus, veiny dock (pl. XXXI, 228).
      - *10b.* Achenes about 2¾ mm. long, 1¾ mm. wide.
        R. altissimus, peach-leaved dock (pl. XXXI, 220).
  - *8b.* Wings of angles not prominent.
    - *11a.* Achenes elliptic, the apex and base about equally pointed; color dark reddish brown, light colored on the angles; length 2–2½ mm., width 1½ mm.
      R. acetosa, garden sorrel (pl. XXXI, 218).
    - *11b.* Achenes ovate, the angles not light colored; length about 2 mm., width 1½–1¾ mm.
      R. salicifolius, willow-leaved dock (pl. XXXI, 227).

Key to *Polygonum*

1a. Achenes flattened, dark reddish brown to almost black.
  2a. Achenes orbicular, smooth and glossy.
    3a. Each side with a sunken area in the center; length 2–2½ mm., width 1¾ mm................*P. lapathifolium*, pale ladysthumb (pl. XXX, 215).
    3b. Sides without sunken area in the center.
      4a. Length 2–2½ mm., width 1¾–2 mm.
P. persicaria, ladysthumb (pl. XXX, 216).
      4b. Length 3–3½ mm., width 3 mm.
P. pensylvanicum, bigseed ladysthumb (pl. XXX, 217).
  2b. Achenes ovate, minutely roughened, flattened at the tip and bulged out below; length 2¾–3 mm., width 2¼ mm.; calyx with small resinous dots, fragments of calyx frequently persisting.
P. hydropiper, common smartweed (pl. XXX, 214).
1b. Achenes 3-angled, the three sides more or less unequal, occasional achenes plano-convex.
  5a. Achenes smooth and glossy.
    6a. Light brown; apex and base about equally short-pointed; two sides equal, the third side broader; length 1½–2 mm., width 1¼–1½ mm.
P. argyrocoleon, silversheath knotweed (pl. XXX, 210).
    6b. Black or dark brown; apex abruptly short-pointed, the base pedicellate.
      7a. Broadly elliptic, the sides prominently concave or sunken between the angles; length 2 mm., width 1¼ mm.
P. hydropiperoides, swamp smartweed.
      7b. Ovate, the sides plane, or if concave only slightly so; length 3–3½ mm., width 1¾–2 mm.; calyx with small resinous dots.
P. punctatum, dotted smartweed.
  5b. Achenes not smooth and glossy.
    8a. Minutely roughened, dull black, but smooth and glossy on the angles; length 3–4 mm., width 2½ mm.
P. convolvulus, black bindweed (pl. XXX, 213).
    8b. Coarsely roughened by small tubercles; dull dark brown but smooth and lustrous on the angles, angles often lie off-center in lower portion, length 2½ mm., width 1½–2 mm...*P. aviculare*, knotweed (pl. XXX, 211).

CHENOPODIACEAE—GOOSEFOOT FAMILY

Numerous species of the goosefoot family are common field weeds. The seeds in this family are mostly flattened, circular or obovate in shape, and the surface is variously configured. (See illustrations in U.S. Department of Agriculture Handbook No. 30, figs. 246–266.) With some exceptions, the embryo is coiled around an abundant endosperm, either in the form of a ring or a horseshoe shape.

Two species are classed as noxious weeds: *Salsola kali* var. *tenuifolia*, Russian thistle, and *Halogeton glomeratus*, halogeton. Halogeton is poisonous to livestock on the ranges, and it is known to have invaded cultivated fields. The seed is reported to have been found in crested wheatgrass and alfalfa. Unlike most species in the goosefoot family, there is no evident endosperm in the seed of these two species. The embryo is spirally coiled into a cone shape or flattened.

The flowers, without petals, are produced in the axils of leaves in fascicles of two or three, or in clusters of 2 to several in halogeton. Two types of flowers are produced on the same plant. In one type the sepals of the calyx are large, rounded, membraneous, and veiny. In the other type the membraneous sepals are lacking.

The seed unit may be the naked spiral, with or without the thin pericarp, or it may be the fruit with its persisting appendages.

Salsola kali var. *tenuifolia*, Russian thistle (pl. XXXI, 230).—The broad, membraneous sepals spread out flat, forming a disk about 7 mm. across and concealing the seed. In the type without the membraneous sepals, one or two of the cone-shaped seeds are firmly locked within hardened calyx and stiff subtending bracts. This structure may be 5–10 mm. long. Such a unit could be mistaken for a piece of sterile stem material. The cone-shaped seeds measure about 2 mm. across the broader end.

Halogeton glomeratus, halogeton (pl. XXXI, 229).—The membraneous sepals are very delicate and spreading and the top of the seed is not concealed. The calyx of the type without membraneous sepals is hardened and only slightly longer than the seed. This structure is about 2 mm. long. Seeds from both forms of flowers are quite flat, apparently never cone-shaped like Russian thistle.

CARYOPHYLLACEAE—PINK FAMILY

There are many field weeds in the pink family. None of the species are agricultural crops.

With the exception of *Dianthus armeria*, deptford pink, the seeds tend to be roundish or oval in outline. The embryo is coiled around the margins of an abundant endosperm (the perisperm of some authors). The tip of the radicle and the tips of the cotyledons terminate on opposite sides of the seed scar or hilum, similar to *Chenopodium* and *Amaranthus*.

Noxious weeds in the pink family are represented in three genera: *Agrostemma*, *Lychnis*, and *Silene*. Over half the States classify *Agrostemma* as a noxious weed. Plants in the *Lychnis* and *Silene* groups are closely related and their seeds are very similar. The species known to occur as field weeds in these two genera are treated in one seed key.

The seeds are black or brown, thick and flattened, the scar lying on the edge. The surface is roughened by tubercles of various types which are arranged in rather definite patterns. The tubercles arise from plate-like areas. The plates are usually gray in color and may have a rather definite shape in some species.

Agrostemma githago, corncockle (pl. XXXI, 231).—Seeds are black and roughly circular in shape. The overall length and width are about equal, 3–3½ mm. The surface is covered with close concentric rows of spine-like tubercles.

Key to *Lychnis* and *Silene*

1a. Elongated plates at the scar collar-like; scar a circular or oval cavity; seeds 1½–2 mm. across or larger; tubercles not arranged in a pattern.

2a. Collar high, completely surrounds the scar cavity or nearly so; tubercles fine, pointed, and so close-spaced the seed appears spiny.
Lychnis dioica, red campion (pl. XXXI, 233).

2b. Collar shorter, surrounds ½–¾ of the scar cavity; tubercles short, blunt, not close-spaced, seed appears smoother than *dioica*.
Lychnis alba, white cockle (pl. XXXI, 232).

1b. Elongated plates at the scar lacking or not collar-like.

3a. Elongated plates at the scar in two pad-like structures like miniature paws, one on each side of the scar cavity.

4a. Tubercles not arranged in a definite pattern on the sides of the seed; seeds thick, with rounded margins.

5a. Seeds 1½ mm. across; tubercles short, blunt, crowded; in edge view tubercles in rows both above and below the scar; the grayish plates roundish or linear-oblong.
Silene noctiflora, night-flowering catchfly (pl. XXXII, 242).

5b. Seeds ½–¾ mm. across; tubercles pointed or blunt, large for the size of the seed; the grayish plates roundish.
Silene anthirrhina, sleepy catchfly (pl. XXXI, 234).

4b. Tubercles arranged in definite patterns on the sides of the seed.

6a. Plates of pads at the scar cavity long and narrow, appearing like fine lines; tubercles pointed, arranged in four or five irregular concentric rows on the sides of the seed; in edge view tubercles not in definite rows above and below the scar; the grayish plates oblong; seeds 1½–2 mm. across, flattened and angular.
Silene cucubalus, bladder campion (pl. XXXII, 240).

6b. Plates of pads at the scar broader and uneven in length; tubercles short and blunt, in three or four prominent concentric rows on the sides of the seed; the grayish plates oblong.

7a. Seeds 1½–2 mm. across, tend to be asymmetrical in shape and obtusely angled; in edge view the tubercles appear crowded, not in distinct rows above or below the scar.
Silene dichotoma, forked catchfly (pl. XXXII, 241).

7b. Seeds 1¼ mm. across, the shape symmetrical; in edge view tubercles stand out sharply in three or four rows above the scar, crowded and much smaller below the scar.
Silene cserei, smooth catchfly (pl. XXXII, 239).

3b. Elongated plates at the scar lacking or much reduced.

8a. Scar a narrow transverse slit.

9a. Tubercles in three prominent concentric rows on the sides of the seed; tubercles fine, pointed, in edge view stand out in three or four sharp, wide-spaced rows above the scar, crowded and much smaller below the scar; seed thick, 1½ mm. across.
Silene conica, conical catchfly (pl. XXXI, 236).

9b. Tubercles not in distinct rows on the sides of the seed.

10a. Seeds 1½ mm. across; tubercles short, wide-spaced and inconspicuous, arising from diamond-shaped plates; in edge view the plates lie transversely on the broad dorsal side.
Silene conoidea, no common name (pl. XXXI, 237).

10b. Seeds 1¾–2 mm. across; tubercles linear or dome-shaped, crowded near the margins on the sides of the seed; in edge view the domes above the scar large, not in rows, much smaller below the scar.
Silene cretica, Cretian catchfly (pl. XXXII, 238).

8b. Scar an oval cavity; seed 1 mm. across, with a depressed crescent-shaped area of elongated cells on the sides below the scar; tubercles short, fine, inconspicuous and crowded, not arranged in a pattern.
Silene gallica, English catchfly (pl. XXXI, 235).

EUPHORBIACEAE—SPURGE FAMILY

Many species of the spurge family may occur as field weeds. Most of the species are rather widely distributed. Only one genus, *Euphorbia*, contains noxious-weed species: Leafy spurge (*E. esula*) and shining spurge (*E. lucida*). Twenty-one species representing the spurge family are described here.

Seeds of the different genera vary greatly in size, shape, and surface configuration. With a few exceptions, the scar is a flattened area at the base. In some species the scar is obscured by a whitish, corky outgrowth, the caruncle.

Many species have a prominent chalaza, which appears as a dark-colored area on the end opposite the scar. A distinctive feature of the seeds in this family is the presence of a prominent raphe. This structure appears as a dark line extending from the scar to the chalazal area at the opposite end.

The embryo of the seed is usually straight and lies in the middle of an abundant endosperm. The radicle is short and the cotyledons are thin and broad.

Six genera are represented in the seed key: *Acalypha*; *Caperonia*; *Croton*; *Crotonopsis*; *Euphorbia*; *Phyllanthus*. Only those species that are likely to occur with crop seeds are included. One crop plant, castorbean (*Ricinus communis*), is in limited cultivation. The seed has the same structural features as described above and the genus is not included in the seed key.

Similar-Appearing Species

Croton.—Several species of *croton* may be found in crop seeds and the seeds are not always distinguishable. Two commonly occurring species are described in the seed key.

Leafy spurge (*Euphorbia esula*) and shining spurge (*E. lucida*).—No features were found whereby the seeds of these two species can be distinguished with certainty. *E. lucida* is reported to set seed infrequently under some conditions.

Key to Euphorbiaceae

1a. Seed enclosed in an indehiscent capsule. Capsule broadly elliptic, with a persisting style remnant and a short-stalked base; both sides of capsule convex, thickest about midway or lower, the tip flattened; light brown or greenish with adhering whitish, fringed scales.
  Length 2–2¼ mm., width 2 mm.—*Crotonopsis linearis*, crotonopsis.
  Length 3–3¼ mm., width 2¼–2½ mm.—*Crotonopsis elliptica*, crotonopsis.

1b. Seed free from the capsule, the raphe a prominent dark line; caruncle present or absent.
  2a. Caruncle linear, persisting, attached from the pointed end to about ⅓ or ½ the length of the seed; chalaza an inconspicuous point; color grayish or dark brown, and spotted with darker brown in *3a*.
    3a. Narrowly obovate, surface so finely reticulate it appears smooth, length 1½ mm., width 1–1¼ mm.
      Acalypha virginica, three-seeded mercury (pl. XLI, 381).
    3b. Broadly obovate, surface with wart-like protuberances; length 2 mm., width 1¾ mm.--*Acalypha ostryaefolia*, hornbeam three-seeded mercury.
  2b. Caruncle cap-like, elevated, easily detached in processing; flattened and persisting in *8a*, and lacking in *5b*.
    4a. Seeds small, 1½ mm. long or less, distinctly angled.
      5a. Seeds quadrangular, slightly flattened in *9b*.
        6a. Length 1 mm., width ½ mm., chalazal area prominent; color pale grayish brown or reddish brown.
          7a. Each face with three or four obscure, transverse ridges.
            Euphorbia supina (*E. maculata*), spotted spurge (pl. XLII, 389).
          7b. Each face faintly pitted, the pits in no definite pattern.
            Euphorbia serpyllifolia, thymeleaf spurge.
        6b. Length 1½ mm., width ¾–1 mm., chalaza obscure except in *9b*.
          8a. Surface pitted, two faces each with two rows of three or four deep pits, and two faces each with a deep longitudinal depression; color light gray, the pits reddish, the cap-like caruncle persisting.
            Euphorbia peplus, petty spurge.
          8b. Surface grooved or lined; color dark brown or grayish.
            9a. The four faces with coarse anastamosing lines; three angles obtuse, smooth and light-colored, the fourth angle grooved or with a dark line.
              Euphorbia nutans (*E. preslii*), nodding spurge (pl. XLI, 388).
            9b. The four faces each with four or five deep, transverse grooves; slightly flattened but distinctly angled.
              Euphorbia falcata, sickle spurge.

5b. Seeds wedge-shaped, without caruncle, the two sides plane, the back side strongly arched; chalazal point and raphe somewhat obscure; color reddish brown.

10a. Surface with longitudinal rows of minute dark brown tubercles; length 1 mm., width ½ mm.
Phyllanthus caroliniensis, Carolina phyllanthus.

10b. Surface with minute transverse lines and scattered black-tipped tubercles; length 1½ mm., width 1¼ mm.
Phyllanthus polygonoides, knotweed phyllanthus.

4b. Seeds large, 2–4 mm. long, neither quadrangular nor wedge-shaped.

11a. Seeds lustrous or glossy; broadly oval or orbicular, slightly flattened, both sides convex with a distinct angle around the periphery, chalaza an inconspicuous point; caruncle, if present, cap-like and elevated.

12a. Surface minutely stippled; light brown and sparingly spotted with dark brown or black; length 3 mm., width 2½–2¾ mm.
Croton glandulosus, glandular croton.

12b. Surface slightly roughened by minute broken lines; light reddish brown or dark grayish brown; length 4 mm., width 3¼ mm.
Croton capitatus, woolly croton (pl. XLI, 382).

11b. Seeds dull.

13a. Surface so minutely reticulate it appears smooth: shape cylindrical, broadly so in *14b*; raphe and chalaza prominent; hilar area in nearly vertical position; caruncle mostly persisting.

14a. Light to dark grayish brown, usually with minute, dark brown flecks...*Euphorbia esula*, leafy spurge (pl. XLI, 384).
Euphorbia lucida, shining spurge.

14b. Light gray with bluish cast, not flecked or mottled.
Euphorbia cyparissias, cypress spurge

13b. Surface lightly or heavily roughened; hilar area in oblique position.

15a. Seeds broadly obovate, rotund.

16a. Surface a porous chalky gray, light brown or spotted with dark brown; chalaza prominent; caruncle lacking.

17a. Surface roughened by an irregular network of obscure, coarse lines and pointed elevations......*Euphorbia marginata*, snow-on-the-mountain (pl. XLI, 387).

17b. Surface obscurely pitted, the pits broad and shallow, sometimes appears merely mottled with brown.
Euphorbia corollata, flowering spurge (pl. XLI, 383).

16b. Surface markedly roughened and minutely stippled; caruncle persisting, the raphe prominent.

18a. Dark reddish brown with whitish wart-like protuberances; chalaza a prominent flat area; caruncle round, covers about one-half of the pilar area.
Euphorbia dentata, toothed spurge (pl. XLI, 385).

18b. Dark brown, with a prominent network of thin reticulations; chalaza an inconspicuous point; caruncle kidney-shaped, covers the entire hilar area.
Euphorbia helioscopia, sun spurge (pl. XLI, 386).

15b. Seeds spherical, diameter 2½–2¾ mm., light to dark grayish brown, with adhering scurfy flecks and a minute reticulum of fine lines; chalaza an inconspicuous point; raphe extends about half way around the seed.
Caperonia palustris, caperonia.

CONVOLVULACEAE—MORNING-GLORY FAMILY

Four genera of the morning-glory family are represented in our field-weed flora, the morning-glories and bindweeds (*Ipomoea*, *Convolvulus*, *Jacquemontia*, *Quamoclit*). The genus *Cuscuta*, dodder, is sometimes included in the morning-glory family. It is classified here in the family *Cuscutaceae*.

The fruit is a two- or four-celled capsule with one or two seeds in each cell. The seeds are thick and wedge-shaped. The two inner faces are equal and plane and the outer face is broad and rounded. Endosperm is present but not copious. The radicle is short and the cotyledons are broad and thin and variously folded. In the descriptions that follow, length represents the long axis of the seed and the width is the broadest point of the seed.

Distinguishing the Genera

The seed key is arranged in two sections, based on the character of the scar. The descriptions apply to the species under consideration and not necessarily to all species in a genus.

Section 1 (pp. 109–110).—Scar is horseshoe-shaped, the narrow end toward the base of the seed. The ends of the scar rim are elevated at the narrow base with a channel or groove between the points. (See pl. XLII, 392.) The scar lies in a horizontal position and is fully exposed when the seed is viewed on the ventral side. The seed coats are smooth or nearly so. This section includes species of *Ipomoea* and *Quamoclit.*

Section 2 (p. 110).—Scar is a broad inverted U-shape, the points wide apart (see pl. XLII, 391), or if oval or kidney-shaped the scar lies with its long axis crosswise of the seed (see pl. XLII, 395). The kidney-shaped scar lies with the indentation toward the base of the seed. There is usually a dome-like elevation at the base of the seed scar, apparently never a channel or groove as in *Ipomoea.* The seed coats are variously roughened. This section includes species of *Convolvulus* and *Jacquemontia.*

Key to *Convolvulaceae*

Section 1

Scar is horseshoe-shape, the narrow end toward the base of the seed.

1a. Scar smooth and glabrous or granular.

2a. Seeds flattened dorso-ventrally, with a fringe of long, dense, tawny hairs around the margin; color dull reddish brown; length 6–7 mm., width 5 mm.
Ipomoea pandurata, clustered blue morning-glory (pl. XLII, 394).

2b. Seeds neither flattened nor pubescent on the margins.

3a. Seeds black, horny and polished; lateral sides plane; back side slopes sharply to the apex and base from a prominent bulge at the middle, permitting the seed to stand upright; scar large, smooth, flush with the seed surface; length 5–6 mm., width 5 mm.
Ipomoea lacunosa, small-flowered white morning-glory.

3b. Seeds brownish black, dull, under magnification appear minutely stippled; lateral sides with a fine distinct line around the margins and usually with one or two transverse wrinkles on each side; back side strongly arched, with a shallow depression down the middle; scar granular and glabrous or with a few short, glandular hairs present; length 5–5½ mm., width 4 mm.
Ipomoea purpurea, common morning-glory (pl. XLII, 393).

1b. Scar covered with an erect coarse and stiff, grayish or reddish pubescence.

4a. Seeds slightly flattened dorso-ventrally, with a few strands of long whitish hairs arising at one point on each side near the top, and a border of short whitish hairs around the densely pubescent scar; surface minutely reticulate; length 5–5½ mm., width 4 mm.
Ipomoea pulchella, pink-flowered morning-glory.

4b. Seeds not flattened dorso-ventrally.

5a. Color grayish black, dull; under magnification appears minutely stippled; lateral sides plane and slightly depressed at the outer margins, the inner angle is a straight line with a prominent scar notch at the base; back side strongly arched, with a slight depression down the center; length 5–5½ mm., width 3–3½ mm.
Ipomoea hederacea, ivyleaf morning-glory (pl. XLII, 392).

5b. Color dark reddish brown, dull and scurfy; lateral sides plane or with a slight bulge at the center; back side strongly arched, usually with two longitudinal folds; length 4–4½ mm., width 3 mm.
Quamoclit coccinea, small red morning-glory.

Key to *Convolvulaceae*

SECTION 2

Scar is a broad inverted U-shape, the points wide apart, or if oval or kidney-shaped the scar lies with its long axis crosswise of the seed.

1a. Seeds small, length 2¾–3 mm., width 2½ mm.; color light brown, dull; surface roughened by blister-like protuberances; lateral sides plane, the angle between them obtuse; the back side strongly arched, with two longitudinal grooves.
Jacquemontia tamnifolia, hairy morning-glory (pl. XLII, 395).

1b. Seeds large, length 3–6 mm. or more.

2a. Scar kidney-shaped, large and prominent, slightly sunken, lies in an oblique position; color dark reddish brown, dull; surface finely roughened by minute, wide-spaced tubercles, some seeds almost smooth; lateral sides with a large depressed oval area in the middle; back side strongly arched; length 5–6 mm., width 4 mm.
Convolvulus sepium, hedge bindweed (pl. XLII, 391).

2b. Scar oval, very small and sunken, lies in a vertical position.

3a. Color light grayish brown, dull, scurfy; shape mostly obovate with a bluntly pointed base; surface roughened by copious, blunt tubercles or short, thick, wavy lines; lateral sides plane; back side strongly arched toward the top and sloping downward to a pointed base; length 4–4½ mm., width 2½–3 mm.
Convolvulus arvensis, field bindweed (pl. XLII, 390).

3b. Color black, dull; shape outlines various.

4a. Surface very finely roughened by irregular broken lines, suggesting a reticulum.

5a. Broadly obovate, lateral sides plane and slightly depressed around the margins; back side strongly arched, the median depression faint or lacking; length 4 mm., width 3 mm.
Convolvulus incanus, hoary bindweed.

5b. Broadly oval; lateral sides plane, the angle between them obtuse, the outer angles sharply defined; back side strongly arched, with a distinct depression down the middle; length 3 mm., width 3 mm. or less------------*Convolvulus hermannioides*, auricled bindweed.

4b. Surface roughened with short, irregular, sharply etched lines or pointed tubercles; broadly oval or ovate; lateral sides plane, the angle between them obtuse; back side strongly arched, the median depression lacking; length 4½–5 mm., width 3½ mm.
Convolvulus occidentalis, western morning-glory

CUSCUTACEAE—DODDER FAMILY

Six species of dodder, *Cuscuta*, occur commonly with crop seeds. Seeds of these species are readily distinguished, but since all species of dodder are classed as noxious weeds in all States but Alaska it is not usually necessary to identify a seed as to species.

The plant is parasitic and the seed contains a rather scanty endosperm with a spirally coiled embryo without cotyledons. (See pl. XLII, 400.) Some seeds of dodder are sterile and it is sometimes necessary to section a seed to determine if it contains an embryo.

The illustrations show the comparative size and shape of the seeds, the character of the scar area, and the texture of the seed coat. These features serve to distinguish the different species. Five of the six species of *Cuscuta* are illustrated:

C. epilinum, flax dodder (pl. XLII, 396).

C. epithymum, clover dodder (pl. XLII, 397).

C. indecora, large-seeded alfalfa dodder (pl. XLII, 398).
C. pentagona, field dodder (pl. XLII, 399).
C. planiflora, small-seeded alfalfa dodder.
C. racemosa var. *chiliana*, Chilean dodder (pl. XLII, 400).

SOLANACEAE—NIGHTSHADE FAMILY

The fruit of the nightshade family may be a fleshy or juicy berry as in the ground-cherries, or it may be a dry capsule as in jimsonweed. In general, the seeds are orbicular, oval or ovate in outline. They are more or less flattened and may be thick or thin.

In some species the seeds are tightly compressed in the spherical berry and some of the seeds may be bent or misshapen. The seed surface may be smooth or variously configured with a reticulum, broken lines or pits. The embryo is curved within an abundant endosperm. (See pl. XLII, 405.)

The nightshade family includes a few important field and vegetable crops, such as tobacco, tomato, pepper, and eggplant, but most of the species are weeds. Two genera include noxious weeds: The nightshades (*Solanum*), and ground-cherries (*Physalis*). Sixteen species in the genera *Solanum*, *Physalis* and *Datura* are described in the seed key.

Structural Features

Shape and size of seeds.—As already pointed out, seeds from a spherical berry may be variously bent or misshapen. The lengths stated in the seed key represent the long axis of the seed, without regard to the position of the seed scar. The width represents the widest point across the seed. All measurements should be considered as approximate only. Because of the natural variation in the length-width proportions, more exact measurements would be of little value.

The seed scar.—The seed scar may be a circular or triangular cavity (pl. XLII, 408), or it may be a narrow wedge-shaped slit (pl. XLII, 405). The width of the opening in the slit varies somewhat with the maturity or age of the seed. In some species the slit remains closed, as if pinched together (pl. XLII, 407).

Seed color.—Seed color depends somewhat on the origin of the seed or on the maturity of the berry or capsule at time of harvest. Occasionally a seed that is normally black may be almost white. Seeds of *Solanum carolinense* are almost orange yellow if harvested before the berry is fully matured, while those removed from the dry and shriveled berry are pale yellow or whitish. The pale yellow type is most commonly found in crop seeds.

Surface configuration.—A magnification of 6 $\times$ or 7 $\times$ is sufficient for viewing the gross features. Seeds of some species are translucent and a higher magnification tends to confuse certain details.

Similar-Appearing Species

Solanum carolinense, horsenettle, and *S. torreyi*, Torrey's nightshade, differ mainly in size. Individual seeds that might be outside their normal size range would be difficult to distinguish. Horsenettle is the more widely destributed of the two species and would be more likely to occur in cultivated fields.

Because of insufficient material for study, it has not been determined whether seeds of *Solanum nigrum* and *S. nodiflorum* are distinguishable. *S. nigrum* is an eastern species and *S. nodiflorum* is a western species.

Also because of insufficient material, it has not been determined whether seeds of *Solanum villosum* and *S. sarachoides* can be distinguished. No constant features have been observed whereby seeds of *Physalis heterophylla* and *P. lanceolata* can be distinguished.

Key to *Solanum, Physalis, Datura*

1a. Surface configuration a definite reticulum.
  2a. Seeds black, thick.
    3a. Reticulum thin, the interspaces small deep pits; surface variously wrinkled or indented; hilum a circular cavity; length 2½ mm.
      Solanum rostratum, buffalo-bur (pl. XLII, 408).
    3b. Reticulum thick, the interspaces shallow; surface with large dimple-like depressions and minute stipples; hilum triangular; length 2–3½ mm.
      Datura stramonium, jimsonweed (pl. XLII, 401).
  2b. Seeds yellow or reddish, flat.
    4a. Lines of reticulum thin.
      5a. Surface deeply pitted, like a honeycomb; seed deeply indented on the margins; hilum a closed slit.
        Physalis lobata, purple-flowered groundcherry (pl. XLII, 402).
      5b. Lines of reticulum usually with white margins; interspaces very small and comparatively deep.
        6a. Seeds flattened at the hilar end as if pinched together.
          7a. Seeds ovate, the tip usually elongated; length 1–1½ mm., width 1–1¼ mm.
            Solanum nigrum, black nightshade (pl. XLII, 407).
            Solanum nodiflorum, small-flowered nightshade.
          7b. Seeds oval, or if ovate short-pointed; length 2–2½ mm., width 1¾ mm. or more.
            Solanum triflorum, cutleaf nightshade (pl. XLIII, 409).
        6b. Seeds not flattened at the hilar end, the hilum a narrow silt; length 1½–1¾ mm., width 1½ mm.____*Solanum villosum*, hairy nightshade.
            Solanum sarachoides, nightshade.
    4b. Lines of the reticulum wavy, thick and translucent.
      8a. Seeds pale yellow.
        9a. Seeds lens-shaped, length and width about equal, 1½–1¾ mm.; hilum ovate, with a smooth rim____*Physalis pubescens*, husk tomato.
        9b. Seeds broadly ovate, length 1¾–2 mm., width 1½ mm.; hilum a closed slit.
          Physalis longifolia, perennial groundcherry (pl. XLII, 403).
      8b. Seeds reddish or deep yellow.
        10a. Seeds lens-shaped, reddish; length and width about equal, 1½–2 mm.; hilum linear-ovate
          Physalis virginiana, Virginia groundcherry.
        10b. Seeds broadly oval, length 2–2½ mm., width 1¾ – 2 mm.; hilum linear-ovate________*Physalis heterophylla*, clammy groundcherry.
          Physalis lanceolata, prairie groundcherry.

1b. Surface configuration various, reticulum obscure or lacking.
  11a. Seeds large, roundish, length and width about equal, 3–4 mm. or more.
    12a. Surface essentially smooth, the seed usually warped longitudinally; hilum ovate, with a thick, smooth rim; color light yellowish brown or dark brown.
      Solanum elaeagnifolium, silverleaf nightshade (pl. XLII, 406).
    12b. Surface not smooth, color yellow.
      13a. Surface with an obscure reticulum of broad, transversely striate ridges; hilum a round or oval cavity with a thick, smooth rim.
        Solanum melongena var. *esculentum*, eggplant.
      13b. Surface with minute but distinct wavy lines, often forming an obscure reticulum toward the periphery; hilum linear-ovate; length 3–4 mm. Compare with *14a*________*Solanum torreyi*, Torrey's nightshade.

11b. Seeds markedly smaller, 1½–2½ mm. long, broadly oval or ovate; pale to deep yellow; surface roughened but without a definite reticulum.

14a. Surface faintly roughened by short irregular lines, often forming obscure concentric rows toward the periphery; hilum linear-ovate; length 2–2½ mm., width 2 mm. or less.
Solanum carolinense, horsenettle (pl. XLII, 405).

14b. Surface coarsely roughened, translucent; hilum a circular or ovate cavity; length 1¾–2 mm., width 1½ mm.
Physalis subglabrata, taperleaf groundcherry or smooth groundcherry (pl. XLII, 404).

COMPOSITAE—COMPOSITE FAMILY

The inflorescence in plants of the composite family is a flower head consisting of a few to many individual flowers borne on a common receptacle or disc. This structure is sometimes mistaken for a single flower. In some species the flowers around the margin of the disc are different from the others. These are called ray flowers. In some species the seeds produced from the ray flowers are entirely different from those of the disc flowers. (For an example, refer to pl. XLIII, 426.)

The seed unit is an achene, which is an indehiscent, 1-seeded fruit, such as sunflower and thistle. The top of the achene is usually depressed around the base of the style. (See pl. XLIII, 415.) In many species there is a fringe of fine bristles or scales (the pappus) around the outer rim of the depression. (See pl. XLIII, 423.) The pappus may consist of several series of graduated bristles or narrow scales. In some species the pappus falls off early or is entirely lacking.

The large embryo is straight and fills the entire seed cavity. The cotyledons lie flatwise against each other, and the radicle and hypocotyl are short. There is no evident endosperm.

It should be borne in mind that the appearance of the achenes may have been altered in processing. Thus, achenes that are normally dull may have acquired a luster, or some of the pubescence or pappus may have been removed .

The composite family is represented by two groups in the seed keys: The centaureas or star-thistles, with 14 species, and the thistles (including the so-called thistles) with 12 species. Five genera are included in the thistle group: *Carduus*, *Cirsium*, *Cnicus*, *Onopordum*, *Silybum*. Both the centaurea and thistle groups include important noxious-weed species.

Key to *Centaurea*

1a. Pappus lacking, achenes dull or slightly lustrous in *americana*.

2a. Scar at the base of achene or nearly so, neither notched nor indented.
Achenes whitish, with faint longitudinal lines or grooves; broadly obovate or oval, length 2–2½ mm., width 1½–2 mm.
C. picris (*C. repens*), Russian knapweed (pl. XLIII, 424).

2b. Scar in notch or indentation on side of achene.

3a. Achenes black with fine raised longitudinal lines and a white circular area at the top; obovate or oval; length 5–6 mm., width 2½–3 mm.
C. americana, American star-thistle.

3b. Achenes faintly or prominently spotted or striped.
4a. Achenes striped.
5a. Achenes olive-green with four prominent light-colored stripes evident to the base and with finer lines between; occasional achenes gray; shape obovate, length 3 mm., width 1¼ mm. (See also *13b.*)
C. diffusa, diffuse star-thistle.
5b. Achenes light yellowish brown or grayish, with lighter very faint longitudinal stripes; length 3 mm., width 1½ mm.
6a. Rim of collar thin and finely serrate, often with remnants of pappus scales; achenes scarcely narrowed toward the blunt base.
C. nigra, black knapweed.
6b. Rim of collar smooth; achenes broadest above the middle, the ventral side straight, the dorsal side tapering from about the middle to a narrow base.
C. jacea, brownscale knapweed (pl. XLIII, 421).
4b. Achenes lightly flecked or streaked.
7a. Achenes dark brown with faint light-colored flecks or streaks; shape obovate; scar notch small and shallow; length 2½ mm., width 1½ mm. (See *15b.*)
C. solstitialis, yellow star-thistle (ray fls.) (pl. XLIII, 426).
7b. Achenes ashy gray or with brownish tinge, finely spotted or streaked with dark brown; broadly obovate, the small scar notch usually a distinct angle; length 3 mm., width 1½–2 mm.
C. calcitrapa, purple star-thistle (pl. XLIII, 418).
1b. Pappus present; achenes glossy, except in *scabiosa, maculosa*, and *diffusa*.
8a. Pappus dense, brush-like, the bristles uneven, erect, tawny or reddish brown. Achenes bluish gray, yellowish on the collar and around the scar; shape oblong; scar notch shallow, about half the length of the achene; length 3–3½ mm., width 1½–2 mm.----- *C. cyanus*, cornflower (pl. XLIII, 419).
8b. Pappus spreading, not dense and brush-like; bristles in 2–4 rows, the outer row short and broad, the next row longer, the innermost row short and converging toward the middle and overtopping the style, except in *iberica, maculosa*, and *diffusa*.
9a. Bristles of pappus prominently ciliate, the longest about 4 mm. long; color grayish brown.
Achenes light grayish brown, occasionally sparingly flecked with darker brown; shape mostly oblong, the scar in a deep angle; length 4½–5 mm., width 2 mm.
C. scabiosa, scabious star-thistle (pl. XLIII, 425).
9b. Bristles only minutely pubescent.
10a. Pappus black, the longest bristles 6–7 mm., stiff and spreading; the innermost scales short and broad with fringed tips and converging toward the middle; achenes 5 mm. long, 2–2½ mm. wide.
Achenes brown or heavily streaked with dark brown, yellowish on the collar and around the scar and usually with a light-colored line down each side; shape oblong, the scar in a distinct angle.
C. sulphurea, sulphur-colored Sicilian thistle (formerly *C. sicula*).
10b. Pappus white, sparse and delicate, the longest bristles about 4 mm., shorter in *maculosa* and *diffusa*; achenes 3 mm. long, 1½ mm. wide; rim of collar finely serrate.
11a. Scar notch hook-like, prominent.
Achenes grayish green with faint light-colored stripes evident to the top of the scar notch; shape oblong.
C. melitensis, Malta star-thistle (pl. XLIII, 423).
11b. Scar notch an indentation or distinct angle, not hook-like.
12a. Achenes olive-green with four prominent light-colored stripes evident to the base and with finer lines between, dull or slightly lustrous, shape obovate; bristles of pappus in two or three rows.
13a. The longest pappus bristles 1½–2 mm., the innermost row short, slender, erect, or if converging, scarcely exceeding the style---------- *C. maculosa*, spotted knapweed (pl. XLIII, 422).
13b. The longest pappus bristles about 1 mm. long, the short innermost row lacking. (See also *5a.*)
C. diffusa, diffuse star-thistle (disk fls.).

12b. Achenes not green.

14a. Achenes grayish, variously streaked with brown; shape oblong.

15a. Innermost row of pappus scales brown or whitish, short and broad at the base, wide-spaced and erect, or if converging toward the middle do not overtop the style.
C. iberica, Iberian star-thistle (pl. XLIII, 420).

15b. Innermost row of pappus bristles slender, numerous, converging toward the middle and overtopping the style. See also *7a.*
C. solstitialis, yellow star-thistle (disk fls.) (pl. XLIII, 426).

14b. Achenes dark gray with numerous yellowish stripes evident to top of scar, yellowish on collar and around the scar; shape broadly elliptic.
Innermost row of bristles short, slender, converging toward the middle and about equal to the style; scar notch short, usually a distinct angle----*C. diluta*, pale-flowered centaurea.

COMPOSITAE

The Thistle Group

The descriptions in the seed key are based on fully matured achenes. Some achenes present in crop seed may have been harvested before fully matured and such seeds may lack important distinguishing features. Achenes of some species of *Carduus* and *Cirsium* are quite similar and when not fully matured some seeds may be indistinguishable. This is especially true of *Carduus crispus*, *nutans*, *acanthoides*, and *macrocephalus*. Achenes from the ray and disc flowers in a single head may be entirely different in some species. Refer to *Carduus pycnocephalus* for an example (pl. XLIII, 415).

Key to Thistles and So-Called Thistles

(*Carduus*; *Cirsium*; *Cnicus*; *Onopordum*; *Silybum*)

1a. Achenes coarsely ribbed or wrinkled.

2a. Achenes cylindrical, coarsely ribbed, dark brown; collar of 10 short horny teeth; pappus of two rows of stiff bristles, the outer row, 9–10 mm. long, arises from the sinuses of the collar, the inner row about 2 mm. long; achenes 8 mm. long or longer, 3 mm. wide------*Cnicus benedictus*, blessed thistle.

2b. Achenes obovate, with four longitudinal ribs and coarse transverse wrinkles, grayish brown and variously streaked with black; apical collar lacking; achenes 5 mm. long, 2½ mm. wide--*Onopordum acanthium*, Scotch thistle.

1b. Achenes not coarsely ribbed or wrinkled, somewhat flattened.

3a. Achenes essentially smooth.

4a. Achenes small, length about 3 mm., width 1 mm.; oblong or slightly curved, usually obscurely 4-angled toward the base, the style slender, not lobed; color dull brownish.
Cirsium arvense, Canada thistle (pl. XLIII, 416).

4b. Achenes markedly larger, mostly 4–6 mm. long.

5a. Glossy, light gray with numerous longitudinal brownish lines; achenes oblong, the brownish collar in a horizontal position; style slender, 5-lobed; achenes 4½ mm. long, 2 mm. wide. (See also *9a.*)
Carduus pycnocephalus, Italian thistle (pl. XLIII, 415).

5b. Dull or slightly lustrous; light grayish or yellowish brown, nearly black in *7b*, and variously streaked or splashed with purple or black; achenes strongly gibbous, the yellow collar in an oblique position.

6a. Style slender, inconspicuous; achenes faintly but copiously streaked with black or dark purple; length 3½–4 mm., width 1½–1¾ mm.
Cirsium vulgare (*C. lanceolatum*), bull thistle (pl. XLIII, 417).

6b. Style stout, conical or cap-like, obscurely lobed.

7a. Achenes light yellowish brown or faintly streaked with dark purple, a wide band of deep yellow below the collar; length 4½–5 mm., width 1¾–2 mm ------ *Cirsium altissimum*, tall thistle. *Cirsium discolor*, field thistle.

7b. Achenes copiously streaked and splashed with black, or almost entirely black; collar narrow on the straight edge of the achene and much wider on the curved side; length 6 mm., width 3–3½ mm ---------------------- *Silybum marianum*, milk thistle.

3b. Achenes with a distinct or faintly pebbled surface.

8a. Surface dull or only slightly lustrous.

9a. Achenes broadly obovate, grayish brown and copiously spotted with minute dark brown flecks; collar reduced to a minute dark brown rim, the style a short point; length 4 mm., width 2 mm. (See also *5a.*) *Carduus pycnocephalus*, Italian thistle (pl. XLIII, 415).

9b. Achenes oblong or slightly curved, pale grayish or yellowish brown.

10a. Style slender, scarcely exceeding the yellowish collar; surface with minute longitudinal grooves or faint lines; achenes 3–3½ mm. long, 1½ mm. wide. *Carduus crispus*, curly thistle (pl. XLIII, 412).

10b. Style stout, knob-like, obscurely lobed; collar not yellow; surface with numerous longitudinal darker brown lines, or occasionally devoid of lines; achenes 3 mm. long, 1¼ mm. wide. *Carduus acanthoides*, plumeless thistle (pl. XLIII, 413).

8b. Surface glossy, yellowish brown and distinctly pebbled; oblong or slightly curved.

11a. Surface with obscure alternating yellowish and brown stripes; style stout, cone-shaped, 5-lobed at the base; achenes 3½ mm. long, 1½ mm. wide -------------------- *Carduus nutans*, musk thistle.

11b. Surface faintly striate or with a faint median line or groove, sometimes devoid of lines; style stout, the apex distinctly 5-lobed; achenes 4 mm. long, 1½ mm. wide. *Carduus macrocephalus*, large-head plumeless thistle (pl. XLIII, 414).

GLOSSARY

Achene.—A small, dry, 1-seeded indehiscent fruit, as in buckwheat.

Acuminate.—Tapering gradually to a long point.

Acute.—Tapering to a short point.

Awn.—A slender bristle at the end or on the back of a structure such as the lemma of grasses.

Bifid.—Two-cleft.

Bract.—A leaflike or scalelike structure subtending a flower or a grass spikelet.

Callus.—A hard calloused area at the base of the floret in many grasses.

Calyx.—The outermost cycle of floral parts, the sepals included.

Caruncle.—An outgrowth at or near the hilum.

Caryopsis.—The fruit of a grass, the grain.

Chalaza.—That part of an ovule where the nucellus and integuments are united; it is always opposite the upper end of the cotyledons; it is evident on the surface of the seeds of many legumes as a distinct spot or elevation.

Ciliate.—Fringed with hairs on the margin, like eyelashes.

Cleistogene.—A seed produced from a flower that has been fertilized without opening.

Composite.—The Composite family, or a member thereof.

Corolla.—The inner cycle of floral parts, including the petals.

Cotyledon.—The first leaf of the embryo, the seed leaf.

Crucifer.—A member of the mustard or Cruciferae family.
Deciduous.—Not persisting, as the awns in certain grasses.
Dehiscent.—Opening by valves or slits to discharge the seeds, as pods of vetch.
Dense.—Crowded.
Disarticulating.—Separating at maturity, as florets separating from the spikelet in many kinds of grasses.
Dorsal.—The back or outer surface of a seed or fruit.
Elliptic.—Having the outline of an ellipse; the greatest width of the seed about midway and sloping to an equally pointed apex and base.
Emarginate.—Having a shallow notch at the apex.
Embryo.—The rudimentary plant in a seed.
Endosperm.—The reserve food stored around the embryo.
Exserted.—Protruding beyond the surrounding organs.
Fascicle.—A dense bundle or cluster, as in buffel grass.
Filiform.—Threadlike.
Floret.—The lemma and palea with included pistil and stamens or the mature grain.
Fruit.—The ripened ovary and associated parts.
Geniculate.—Bent like a knee.
Gibbous.—Humped or swollen on one side.
Glabrous.—Devoid of hairs or pubescence.
Glaucous.—Covered with a bluish or whitish bloom.
Globose.—Shaped like a globe.
Glumes.—The empty bracts at the base of a grass spikelet; may be thin and chaffy or hardened.
Hilum.—The scar on the surface of the seed, the place of detachment of the mature ovule from its base or stalk.
Hirsute.—Pubescent with straight, rather stiff hairs.
Hispid.—Pubescent with stiff or rigid hairs.
Hyaline.—Thin and transparent.
Hypocotyl.—A transition region between the root and stem of the embryo.
Indehiscent.—Fruits which do not open at maturity.
Involute.—Inrolled, both edges rolled in toward the middle.
Keel.—A sharp longitudinal fold like the keel of a boat, such as along the midrib of the lemma in some grasses.
Lance-shaped.—Broadest below the middle, tapering to a long-pointed apex.
Lemma.—One of the chaffy bracts enclosing the grain of a grass floret.
Linear.—Long and narrow with parallel sides.
Loment.—A legume pod constricted between the seeds.
Mass of seed.—Refers to the distribution of the substance of a seed as it affects the seed shape.
Membraneous.—Thin and transparent.
Mucronate.—Tipped with a short, slender point, such as a minute awn or exserted midnerve in grasses.
Micropyle.—A minute opening on the seed surface, through which the pollen tube entered the ovule.
Moniliform.—Resembling a string of beads.
Nerves.—Ribs or veins in the chaffy structures of grass or seed pods.

Noxious-weed seed.—Seeds from any plant considered to be extremely destructive or harmful to agriculture. These seeds are designated by law for State and Federal law enforcement.
Oblique.—Slanting or tilted.
Oblong.—About two to four times longer than wide, the sides approximately parallel.
Obovate.—Inverted egg-shape.
Obtuse.—Blunt or rounded at the tip.
Oleaginous.—Oily.
Orbicular.—Circular in outline.
Oval.—Broadly elliptic.
Ovate.—Egg shape.
Ovule.—The structure which after fertilization develops into a seed.
Palea.—One of the chaffy bracts enclosing the grain of a grass floret.
Papillae.—Minute nipple-like protuberances.
Pappus.—A bristle-like or feathery structure at the top of the achene in the Compositae, as the thistledown of a dandelion.
Pedicellate.—Provided with a stalk or pedicel, not sessile.
Pericarp.—The wall of the mature ovary.
Persisting.—Remaining attached.
Pistil.—The seed-producing organ of the flower.
Pod.—A capsule or a dry seed vessel.
Prickle.—A sharp, pointed emergence.
Pubescent.—Hairy, usually refers to fine, short hairs.
Punctate.—Dotted with minute holes or depressions.
Rachilla.—The axis of a grass spikelet.
Rachis.—The axis of a spike or raceme.
Radicle.—The rudimentary root of an embryo.
Raphe.—A seamlike ridge or furrow, as the ridge between the hilum and chalaza in some legumes.
Ray flowers.—The marginal, petal-like flowers of some composites.
Reticulate.—Having a network of veins or nerves.
Retrorse.—Hairs or spines directed downward or back.
Rotund.—Rounded out, plump.
Scabrous.—Rough to the touch, covered with minute points or very short, stiff hairs.
Seed.—A fertilized, mature ovule (a true seed), or a unit of commercial seed consisting of a true seed with persisting appendages (botanically a fruit).
Sessile.—Without stem or stalk.
Sinus.—The space between the margins of the lemma at the base of a floret, as in *Agropyron.*
Stamen.—A male floral organ which bears pollen grains.
Sterile.—Barren, without any floral parts.
Stigma.—The tip of a pistil, usually sticky or hairy.
Stipitate.—Having a stalk-like base.
Stipples.—Small spots, or microscopic pits that appear like minute spots, as on the surface of *Brassica* species.
Striate.—Marked with slender longitudinal lines or furrows.
Style.—The stalk-like portion of a pistil connecting the stigma and the ovary.
Suture.—A line of splitting or dehiscence.

Tribe.—A group of related genera forming a natural division, as the tribes of grasses.
Trigonous.—Three-sided.
Truncate.—Terminating abruptly as if cut off crosswise.
Tuberculate.—Having small, pimple-like protuberances.
Undulate.—Having shallow, smooth ridges, usually lying crosswise of the seed structure.
Villous.—Bearing long, soft hairs.

SELECTED REFERENCES

There are many excellent botanical textbooks, manuals, and monographs, but few of them include any information on the identification of seeds of the plants. Departments of Agriculture of some States or the State experiment stations publish weed bulletins pertaining to their local floras. Some of these include illustrations of the seeds but usually no descriptions.

The following are a few of the more comprehensive treatments that pertain entirely or in part to the identification of seeds. The list is not intended to be complete. Some of the older publications that are no longer available, and those that treat only a few kinds have not been included.

GRASSES

COLBRY, V. L.
1957. Diagnostic characteristics of the fruits and florets of economic species of North American Sporobolus. Contr. United States Natl. Herb. 34 (1): 1–24.

POHL, R. W.
1954. HOW TO KNOW THE GRASSES. 192 pp., illus. Wm. C. Brown Co., Dubuque, Iowa. [The flowering habit and spikelets of 293 species of the more common and important kinds of grasses are illustrated and described, together with keys and distribution maps. A brief and excellent reference text.]

HITCHCOCK, A. S. (Second edition revised by AGNES CHASE.)
1950. MANUAL OF THE GRASSES OF THE UNITED STATES. 1051 pp., illus. U.S. Dept. Agr., Misc. pub. 200. Supt. of Documents, Washington 25, D.C. [Includes illustrations and descriptions of all grasses known to grow in the continental United States, excluding Alaska. Good illustrations of the spikelets, accompanied by distribution maps.]

JENSEN, LOUISA A., and others.
1957. SEED CHARACTERISTICS OF CERTAIN WILD BARLEY, *Hordeum* spp. Proc. Assn. Official Seed Analysts 1957: 87–91, illus.

MUSIL, ALBINA F.
1942. DISTINGUISHING CHARACTERS OF SEEDS OF FOUR SPECIES OF *Agropyron*. Proc. Assn. Official Seed Analysts 1942: 124–131, illus.

———
1942. DIAGNOSTIC CHARACTERS OF SEED OF THE COMMERCIAL SPECIES OF *Agrostis* . . . Proc. Assn. Official Seed Analysts 1942: 139–151, illus.

———
1949. DISTINGUISHING THE SEEDS OF CULTIVATED FINE-LEAVED FESCUES . . . Proc. Assn. Official Seed Analysts 1949: 93–107, illus.

———
1950. SEEDS OF *Agropyron*, ELEVEN SPECIES OF ECONOMIC INTEREST COMPARED. Proc. Assn. Official Seed Analysts 1950: 59–71, illus.

———
1957. A COMPARISON OF THE SEEDS OF *Sorghum almum*, *S. halepense* AND *S. sudanense*. Proc. Assn. Official Seed Analysts 1957: 80–86, illus.

PEPIN, A., and WRIGHT, W. H.
[n.d.] THE IDENTIFICATION OF FALSE WILD OATS . . . [Pamphlet contains descriptions and illustrations of *Avena sativa*, *A. fatua*, and the fatuoids.] Contr. to AOSA Handbook.[4]

WRIGHT, W. H.
1935. SEEDS OF THE GENUS *Poa*. [Pamphlet contains descriptions and illustrations of four species of *Poa*.] Contr. to AOSA Handbook.[4]

GENERAL SUBJECTS

BEIJERINCK, W.
1947. ZADENATLAS DER NEDERLANDSCHE FLORA. H. Veenman & Zonen, Wageningen. [Seed illustrations, 140 page-plates of 10 figures each. No descriptions.]

BELLUE, MARGARET K.
1949. WEED SEED HANDBOOK. (Reprinted from Weed Seed Handbook Series No. I–VIII, Dept. Agr. Bul., V. 34, 35, 36.) State Dept. Agr., Sacramento, California. [Describes and illustrates the plants and seeds of the primary and secondary noxious weeds of California.]

BROUWER, WALTHER, and STÄHLIN, ADOLF.
1955. HANDBUCH DER SAMENKUNDE. DIG-VERLAGS-GMBH., Frankfurt am Main. [Text in German. About 2,500 seeds described and many illustrated.]

HEINISCH, O.
1955. Samenatlas. 122 pp. illus. [Descriptions in three or four languages; illus. in black and white.]

HYDE, E. O. C.
[n.d.] WEED SEEDS IN AGRICULTURAL SEED BUL. NO. 316, New Zealand Dept. Agr., Palmerston North. [48 pp., illustrations and descriptions of 172 species.]

ISELY, DUANE.
1947. INVESTIGATIONS IN SEED CLASSIFICATION BY FAMILY CHARACTERISTICS. Research Bul. No. 351, Ames, Iowa. 380 pp., illus. [Contains descriptions of family characteristics and eight page-plates showing gross internal morphology of seeds in 17 plant families.]

———
1948. SEED CHARACTERS OF ALFALFA AND CERTAIN SPECIES OF *Medicago*. Contr. to AOSA Handbook.[4] [Descriptions and illustrations of four species of *Medicago*.]

———
1949. SEED CHARACTERISTICS OF WEEDY THISTLES. Contr. to AOSA Handbook.[4] [Descriptions and illustrations of 12 species.]

——— and WRIGHT, W.H.
1954. NOXIOUS WEED SEEDS. Iowa State College Jour. of Science, Ames, Iowa. Vol. 28, No. 4, 521–586, illus. [Seeds of certain noxious weeds are illustrated and described.]

KORSMO, EMIL.
1935. WEED SEEDS. Clydendal Norsk Forlag, Oslo, Norway. 175 pp., illus. 306 species are described and illustrated in color. [Text in Norwegian, German, English. Common names in Norwegian, German, Danish, French, Russian, Italian, Dutch, American, Swedish, and English, for U.S.A., Canada and England.]

MUSIL, ALBINA F.
1948. DISTINGUISHING THE SPECIES OF *Brassica* BY THEIR SEED. U.S. Dept. of Agr., Misc. Pub. No. 643. 35 pp., illus. [Includes species that are being grown in this country, imported for planting or other purposes, or occurring in fields and waste places.]

PIERPOINT, MERLE.
1949. CERTAIN VETCHES AND SIMILAR PLANTS. Contr. to AOSA Handbook.[4] [Pamphlet contains descriptions and illustrations of 20 species.]

Copies are available from the secretary-treasurer of the Association of Official Seed Analysts.

U.S. DEPT. AGR. HANDBOOK NO. 30.
1952. MANUAL FOR TESTING AGRICULTURAL AND VEGETABLE SEEDS. Supt. of Documents, Washington 25, D.C. [Contains 34 page-plates of seed illustrations, 735 species representing 55 families, described in part.]

UNIVERSITY OF ILLINOIS.
1954. WEEDS OF THE NORTH CENTRAL STATES. Univ. of Illinois, Agr. Expt. Sta., Circ. No. 718, Urbana, Ill. 239 pp., illus. [More than 200 species are described and illustrated, together with distribution maps. Keys to the plants were prepared by Duane Isely.]

WRIGHT, W. H.
1940. THE SEEDS OF *Sonchus*. Contr. to AOSA Handbook.[4] [Describes and illustrates three species of *Sonchus*.]

——— [1950–1955] WEED SEEDS. [Consists of eight sections of seed illustrations in color, 36 species in each section, no descriptions. Thirty-eight families are represented.] Obtainable at $1.50 per section from the Queen's Printer, Ottawa, Canada.

[4] Copies are available from the secretary-treasurer of the Association of Official Seed Analysts.

INDEX

Common Plant Names

600705 O—62——9

Botanical Plant Names

U.S. GOVERNMENT PRINTING OFFICE: 1963 O—600705

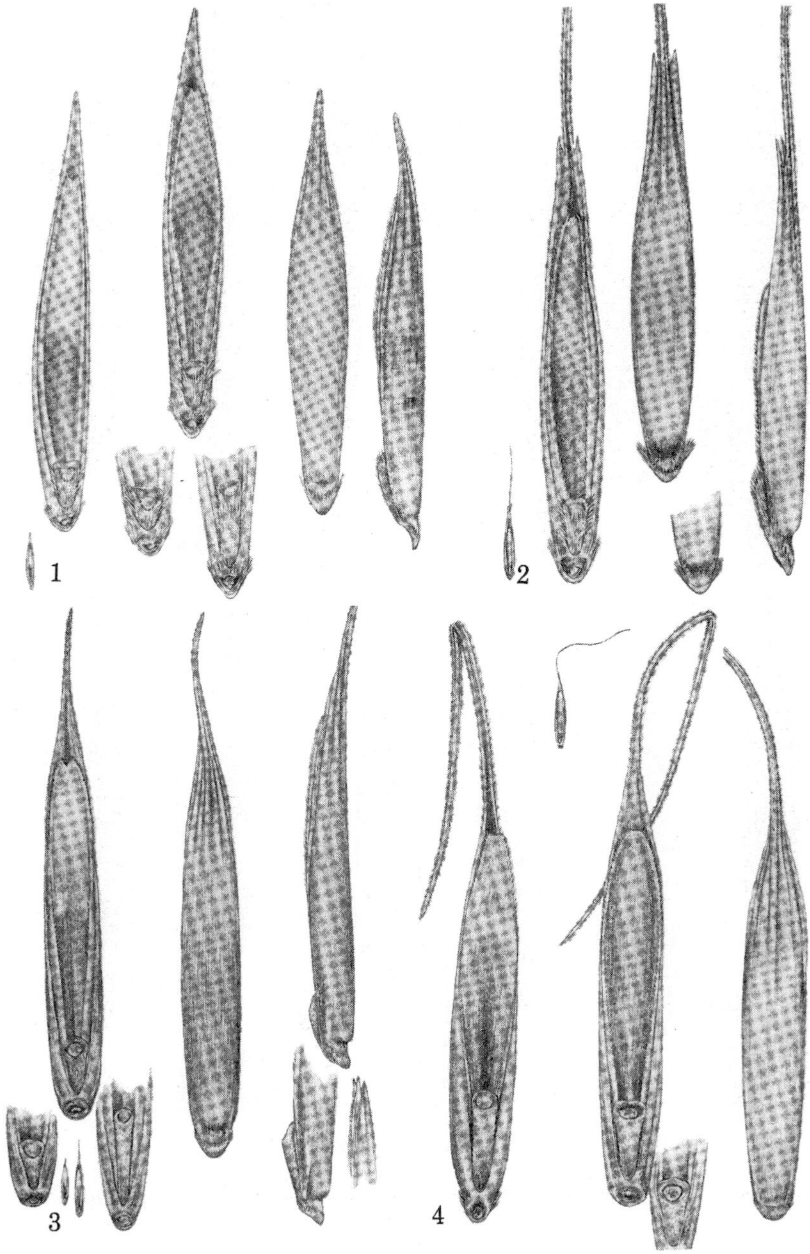

Agropyron (1–4), × 10

1. *A. trachycaulum*, slender wheatgrass
2. *A. subsecundum*, bearded wheatgrass
3. *A. inerme*, beardless wheatgrass
4. *A. spicatum*, bluebunch wheatgrass

600705 O - 63 - 12

PLATE II

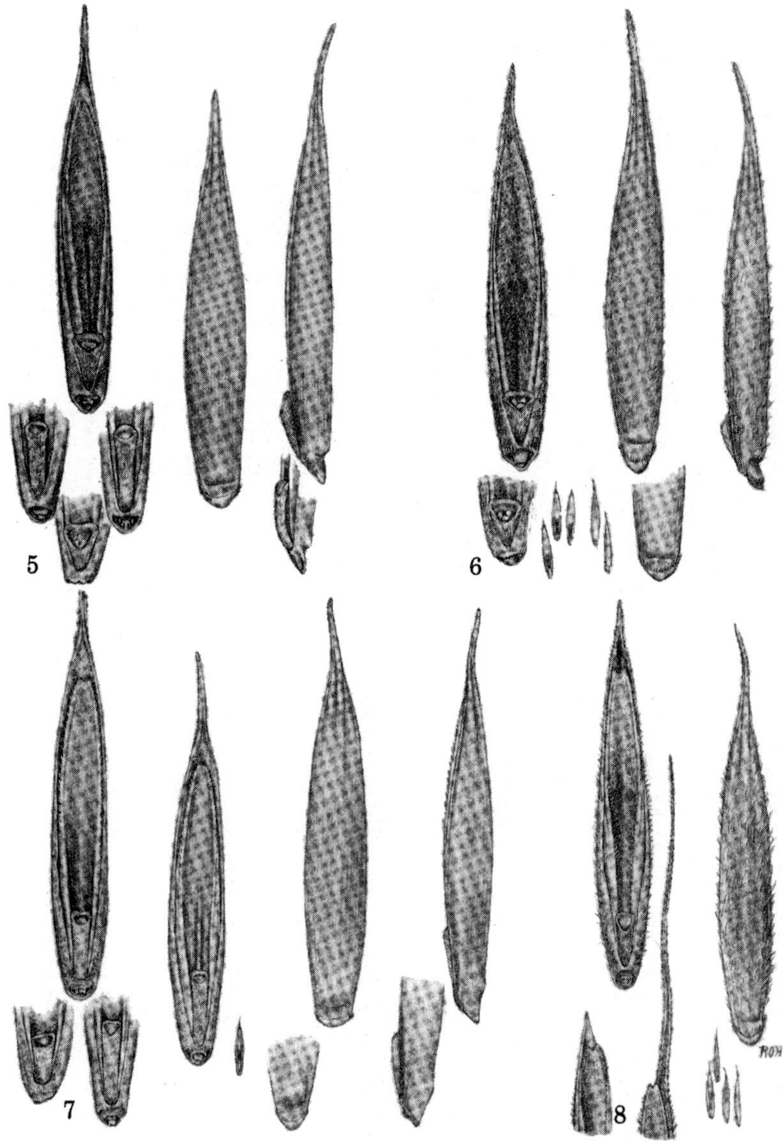

Agropyron (5–8), × 10

5. *A. smithii*, western wheatgrass
6. *A. smithii* var. *molle*, hairy western wheatgrass
7. *A. repens*, quackgrass
8. *A. riparium*, streambank wheatgrass

PLATE III

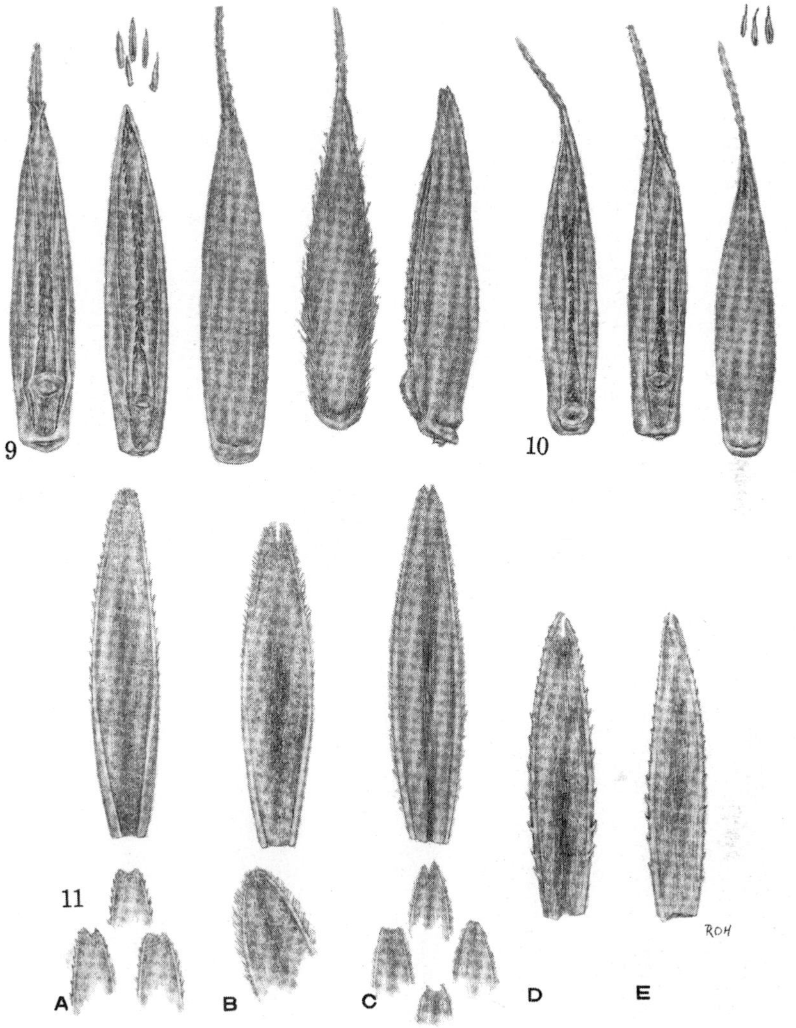

Agropyron (9–11), × 10

9. *A. desertorum*, standard crested wheatgrass
10. *A. cristatum*, Fairway crested wheatgrass
11. Paleas of five species of *Agropyron*: (A) *A. repens*, (B) *A. smithii*, (C) *A. trachycaulum*, (D) *A. desertorum*, (E) *A. cristatum*

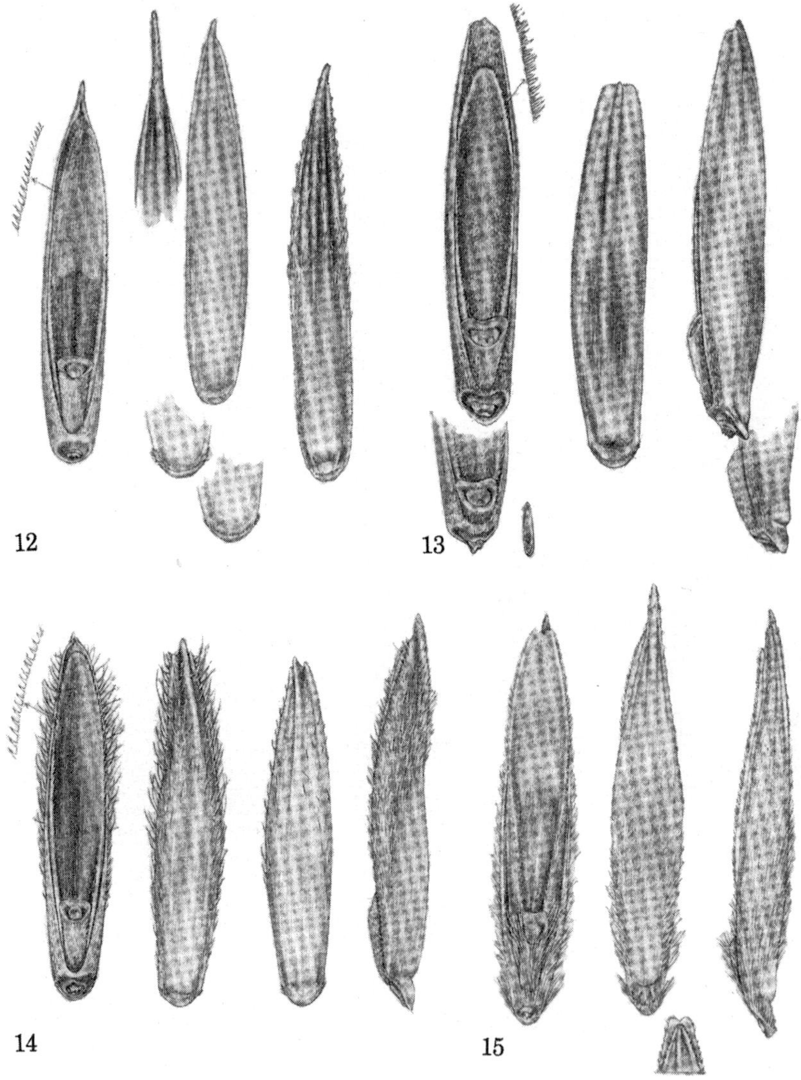

Agropyron (12–15), × 10

12. *A. intermedium*, intermediate wheatgrass
13. *A. elongatum*, tall wheatgrass
14. *A. trichophorum*, hairy intermediate wheatgrass
15. *A. dasystachyum*, thickspike wheatgrass

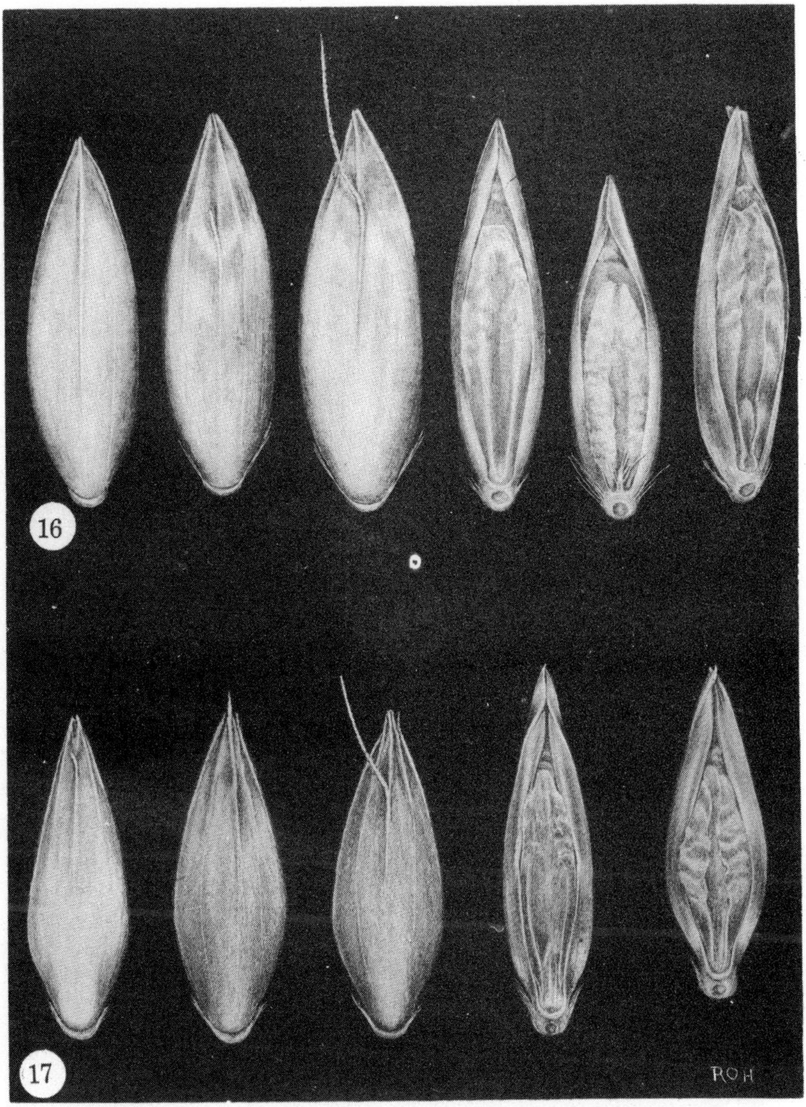

Agrostis (16–17), × 25

16. *A. alba*, redtop

17. *A. palustris*, creeping bentgrass

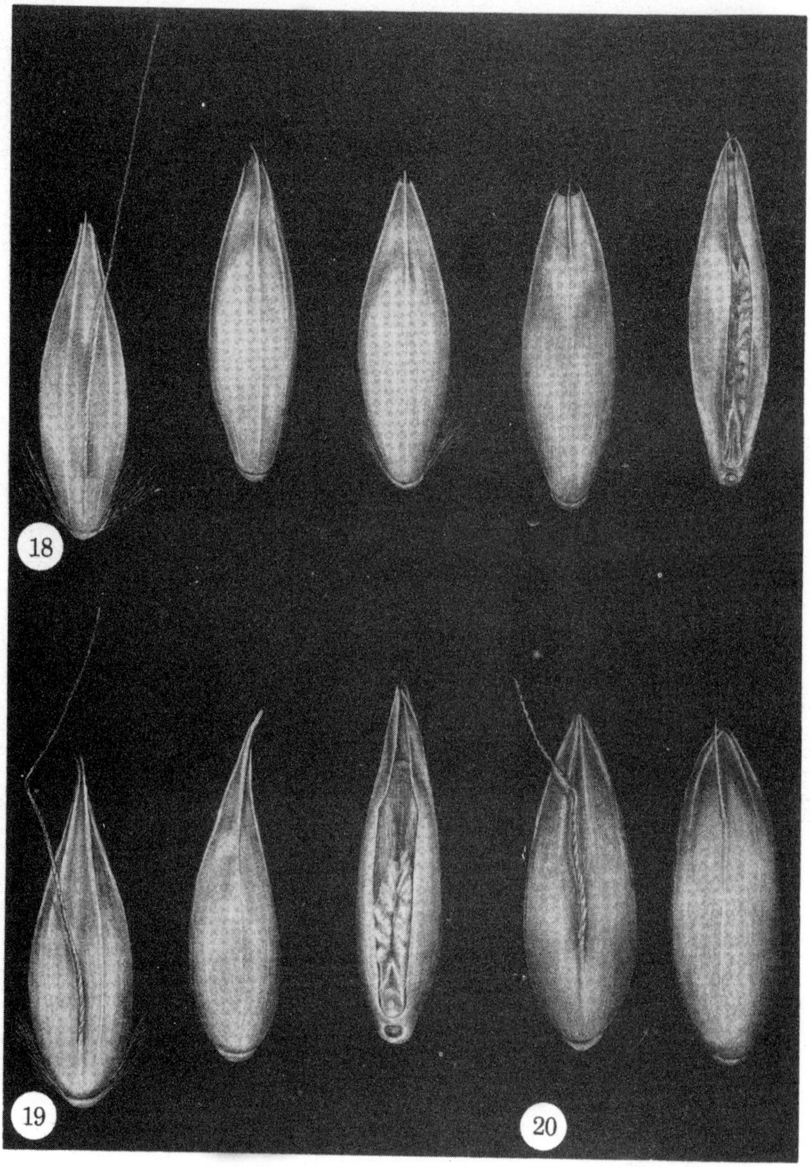

Agrostis (18–20), × 25

18. *A. tenuis* colonial bentgrass var. Highland
19. *A. tenuis*, colonial bentgrass
20. *A. tenuis*, colonial bentgrass var. Astoria

PLATE VII

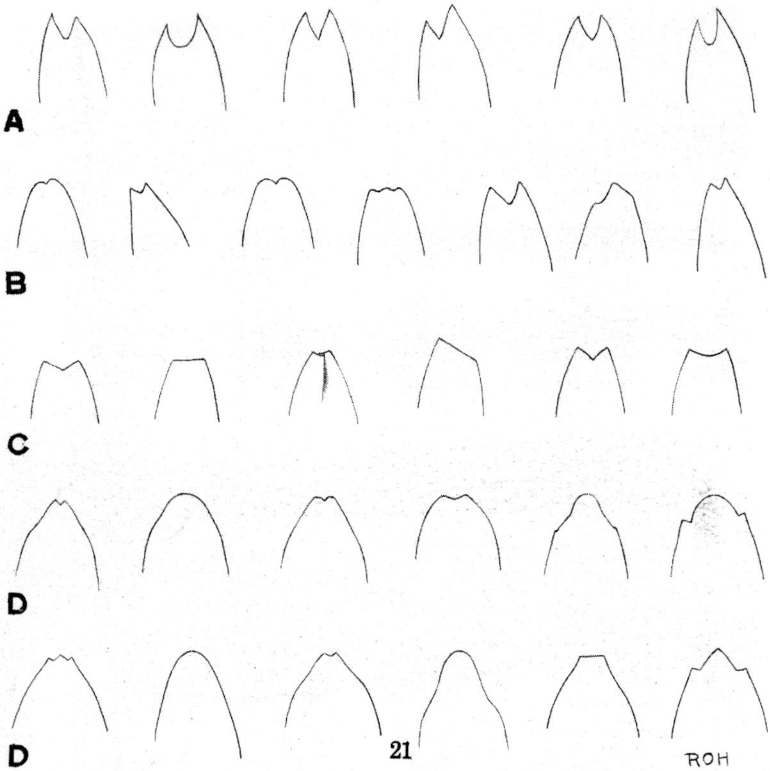

Agrostis (21)

21. Characteristic palea tips of three species and two varieties, greatly enlarged:
(A) *A. tenuis*, colonial bentgrass var. Highland
(B) *A. tenuis*, colonial bentgrass, including the var. Astoria
(C) *A. alba*, redtop
(D) *A. palustris*, creeping bentgrass

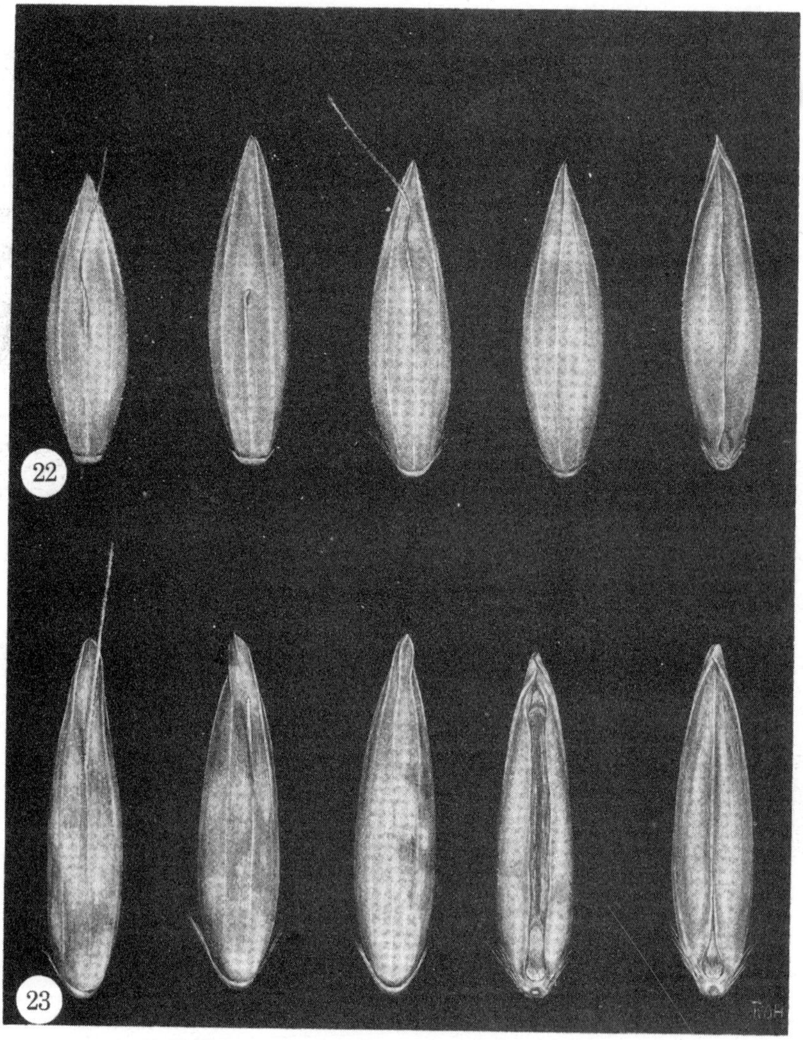

Agrostis (22–23), × 25

22. *A. canina*, velvet bentgrass

23. *A. scabra*, ticklegrass

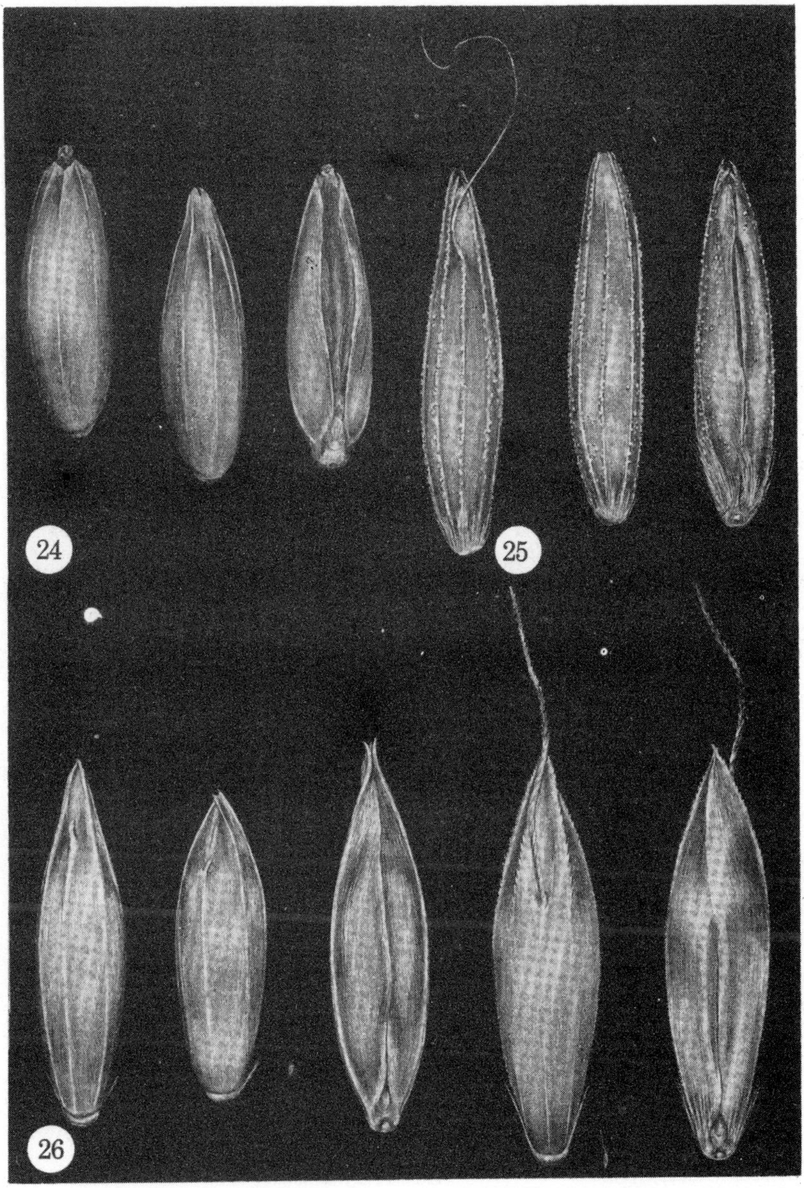

Agrostis (24–26), × 25

24. *A. hiemalis*, winter bentgrass
25. *A. elliottiana*, Elliott bentgrass
26. *A. exarata*, spike bentgrass

27

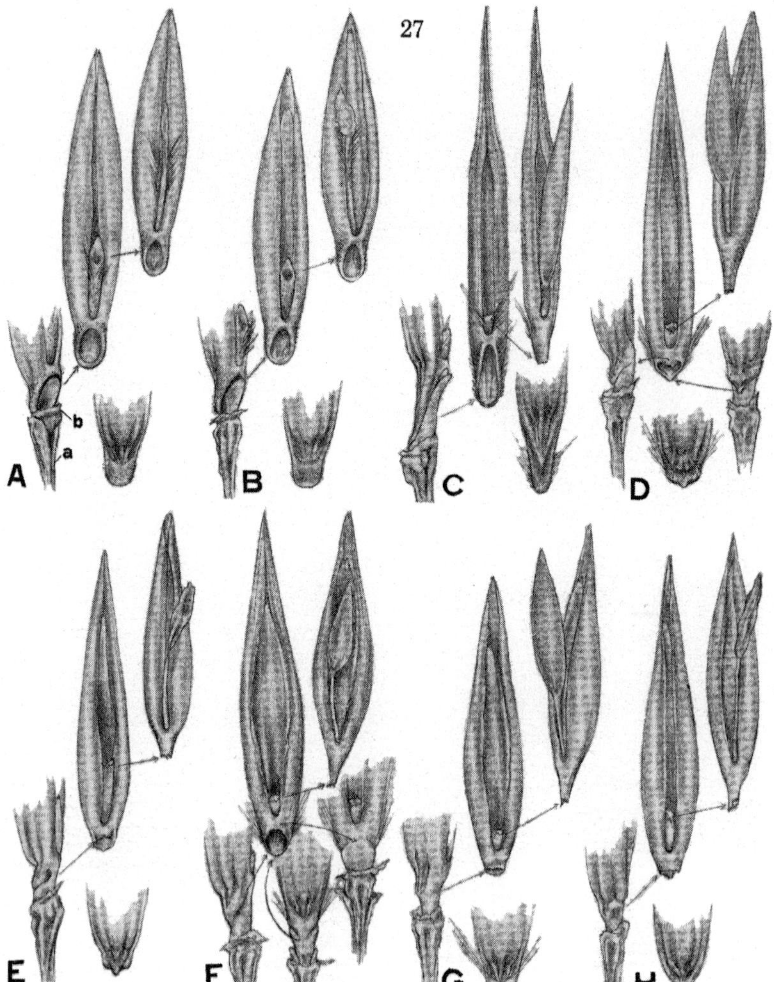

Avena (27)

27. Species of *Avena* illustrating types of abscission of floret in a spikelet, from complete and partial abscission in the wild species, to complete fracture in the cultivated oats. Each figure shows a complete spikelet. The detail at the lower left in each figure shows the position of the basal floret on its pedicel. The florets are somewhat stylized in order to show the points of articulation clearly. Awns and some of the pubescence are omitted.

(A) *A. fatua*, wild oat
(B) Fatuoid (homozygous)
(C) *A. sterilis* var. *maxima*, wild red oat
(D) *A. sterilis* var. *ludoviciana*, wild red oat
(E) Fatuoid (heterozygous)
(F) *A. byzantina*, oat var. Red Rustproof
(G) *A. byzantina*, oat var. Fulghum
(H) *A. sativa*, oat var. Victory

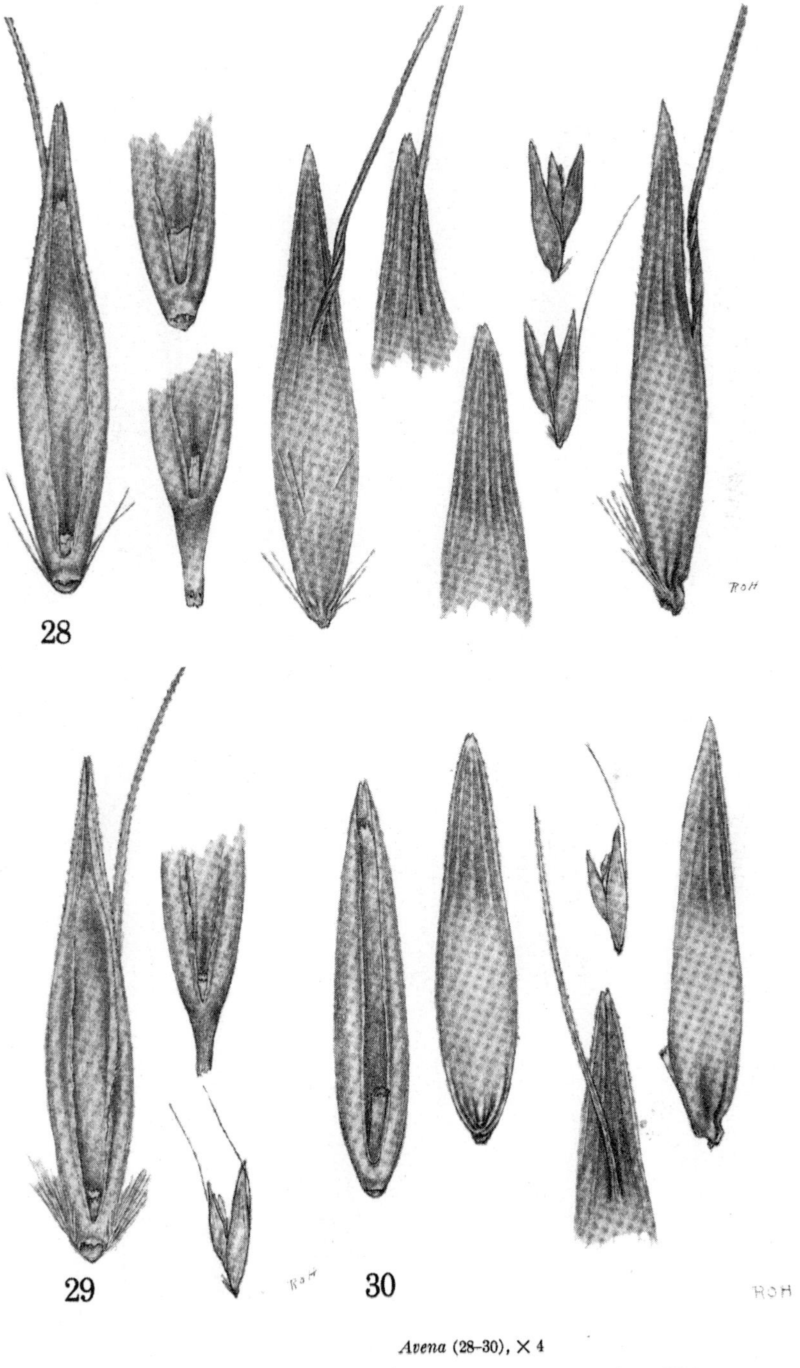

Avena (28–30), × 4

28. *A. byzantina*, oat var. Fulghum
29. *A. byzantina*, oat var. Red Rustproof
30. *A. sativa*, oat var. Victory

PLATE XII

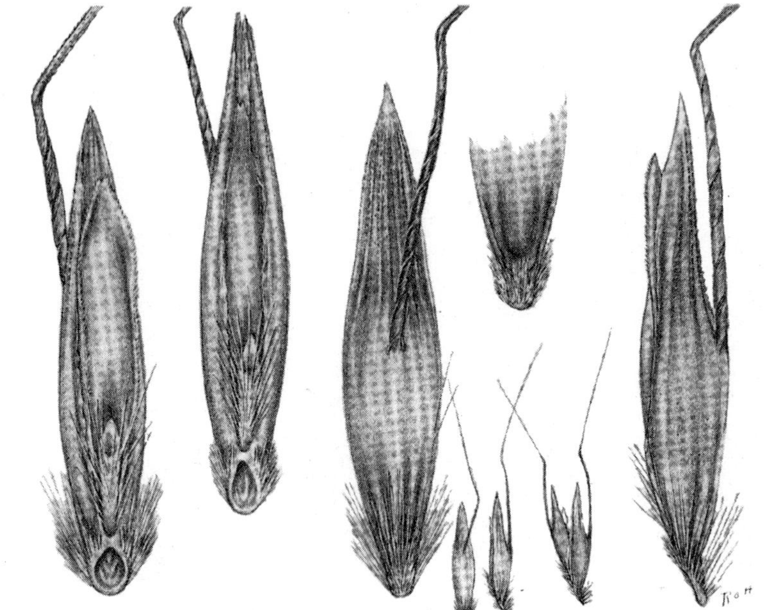

31

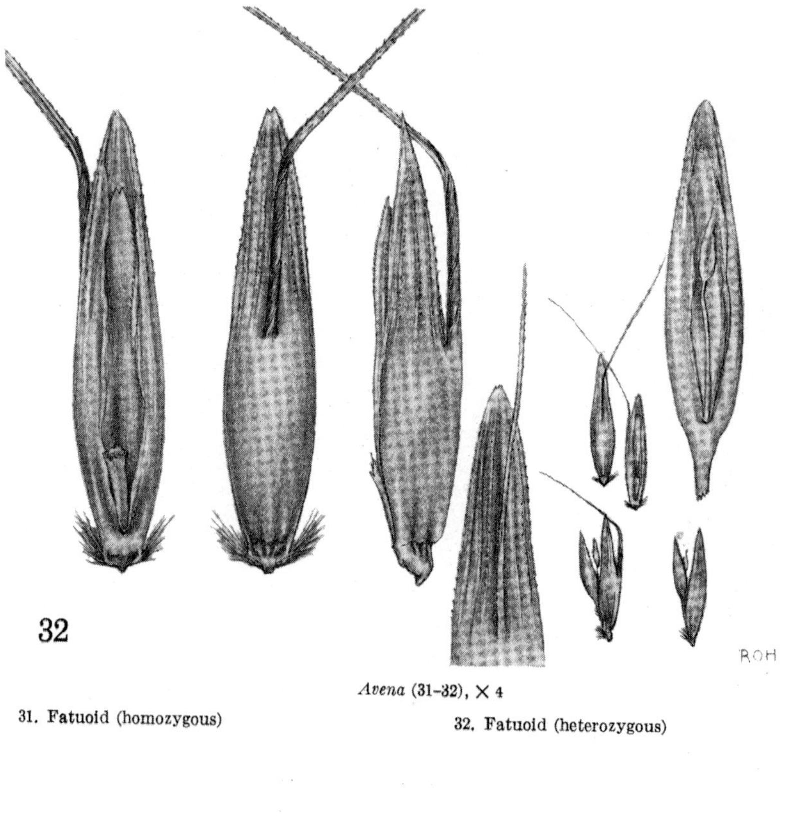

32

Avena (31–32), × 4

31. Fatuoid (homozygous)

32. Fatuoid (heterozygous)

PLATE XIII

33

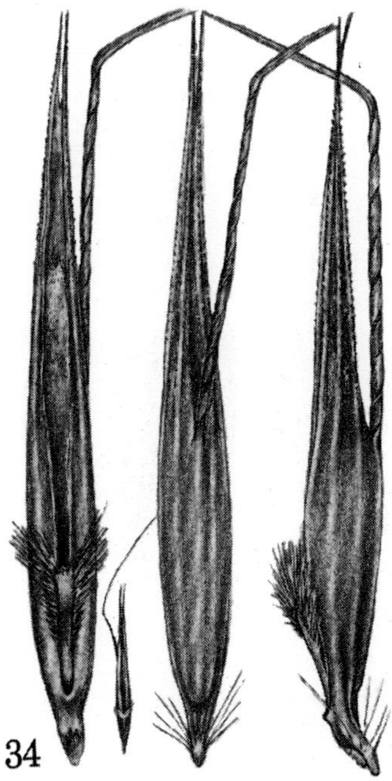

34

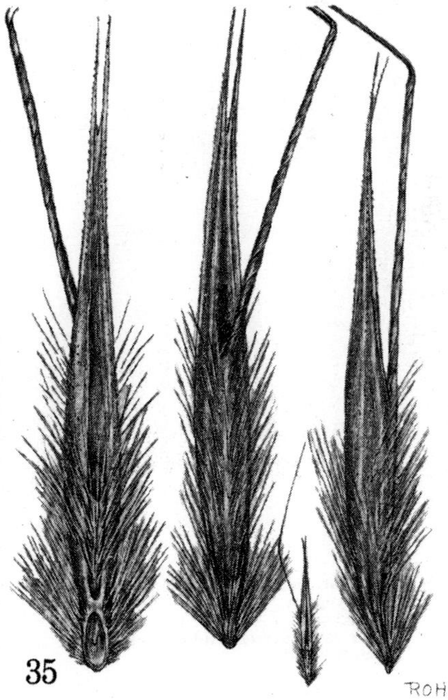

35

Avena (33–35), × 4

33. *A. fatua*, wild oat
34. *A. strigosa*, sand oat
35. *A. barbata*, slender oat

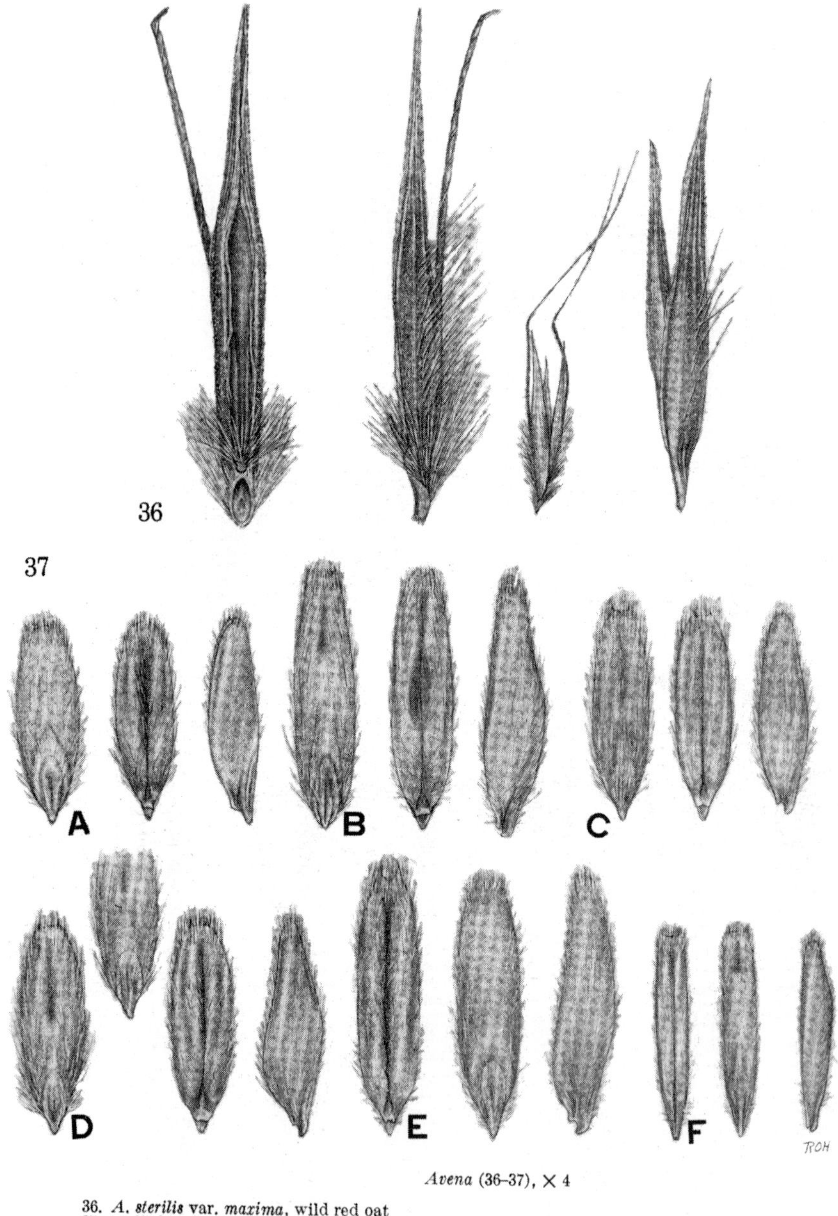

Avena (36–37), × 4

36. *A. sterilis* var. *maxima*, wild red oat
37. Hulled caryopses:
(A) Cultivated oats
(B) *A. sterilis* var. *maxima*, wild red oat
(C) *A. fatua*, wild oat
(D) Fatuoid (homozygous)
(E) Fatuoid (heterozygous)
(F) *A. barbata*, slender oat

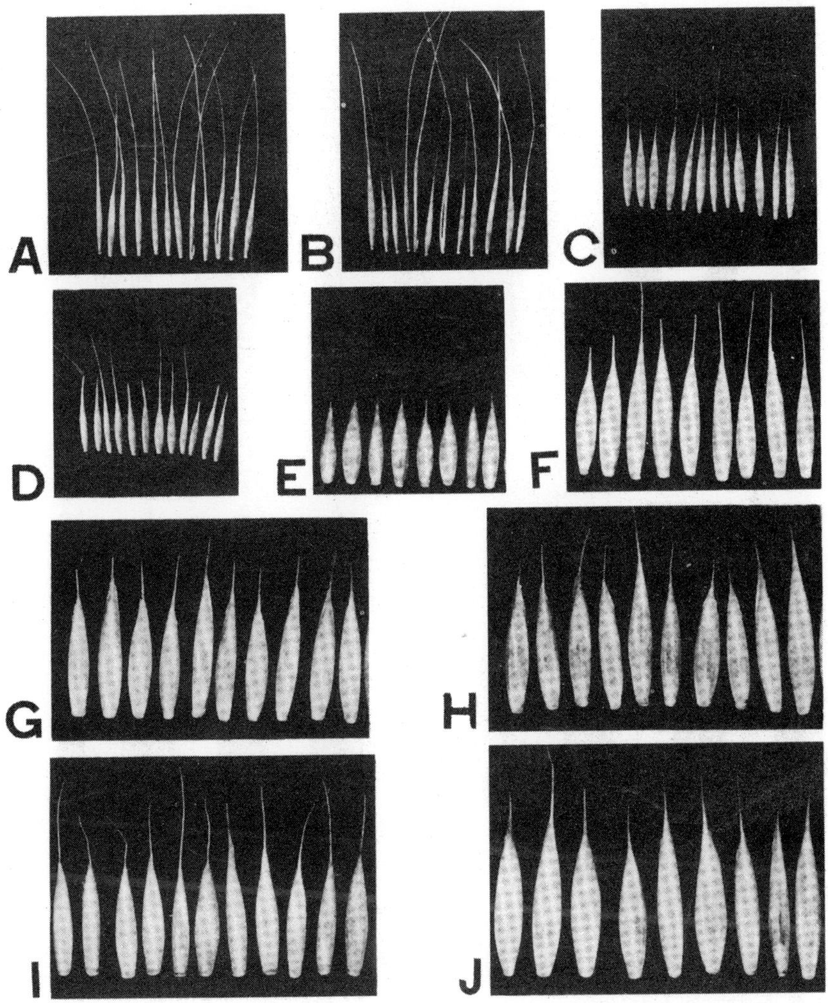

38. Seeds of fine-leaved fescues, 2 × natural size, showing comparative size and shape in dorsal view:

(A) *F. myuros*, rattail fescue
(B) *F. megalura*, foxtail fescue
(C) *F. dertonensis*, big rattail fescue
(D) *F. octoflora*, sixweeks fescue
(E) *F. capillata*, hair fescue
(F) *F. ovina*, sheep fescue
(G) *F. rubra* var. *commutata*, Chewings fescue
(H) *F. rubra*, red fescue var. Illahee
(I) *F. rubra*, commercial red fescue
(J) *F. rubra*, red fescue var. Rainier

PLATE XVI

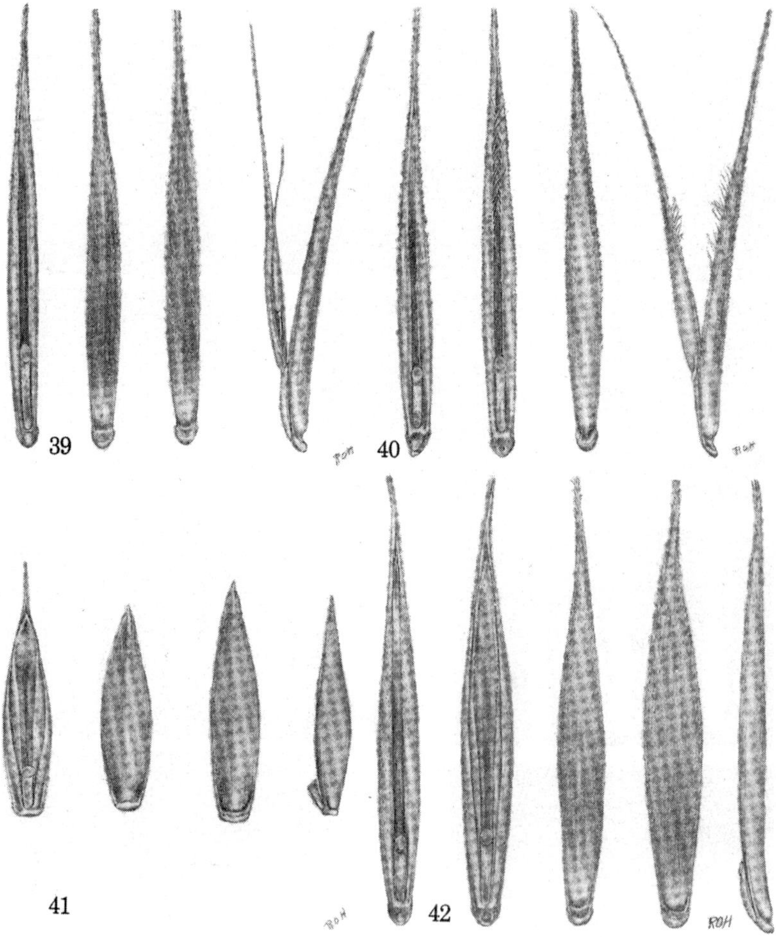

Festuca (39–42), × 9

39. *F myuros*, rattail fescue
40. *F. megalura*, foxtail fescue
41. *F. capillata*, hair fescue
42. *F. dertonensis*, big rattail fescue

PLATE XVII

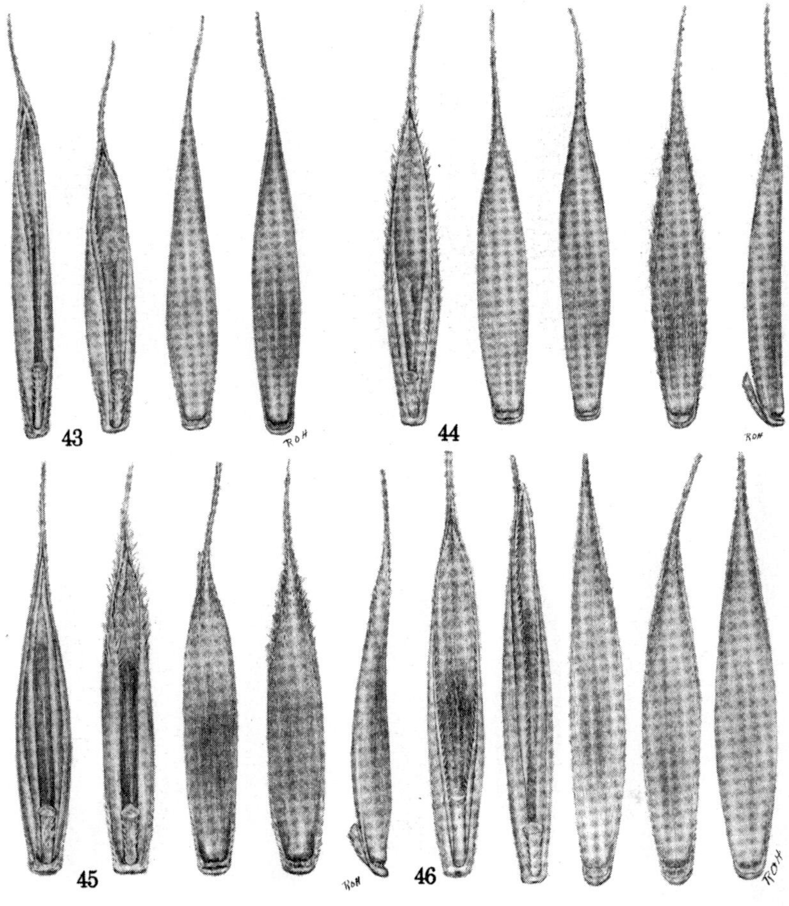

Festuca (43–46), × 9

43. *F. rubra* var. *commutata*, Chewings fescue
44. *F. rubra*, red fescue var. Illahee
45. *F. ovina*, sheep fescue
46. *F. rubra*, red fescue var. Rainier

600705 O - 63 - 13

PLATE XVIII

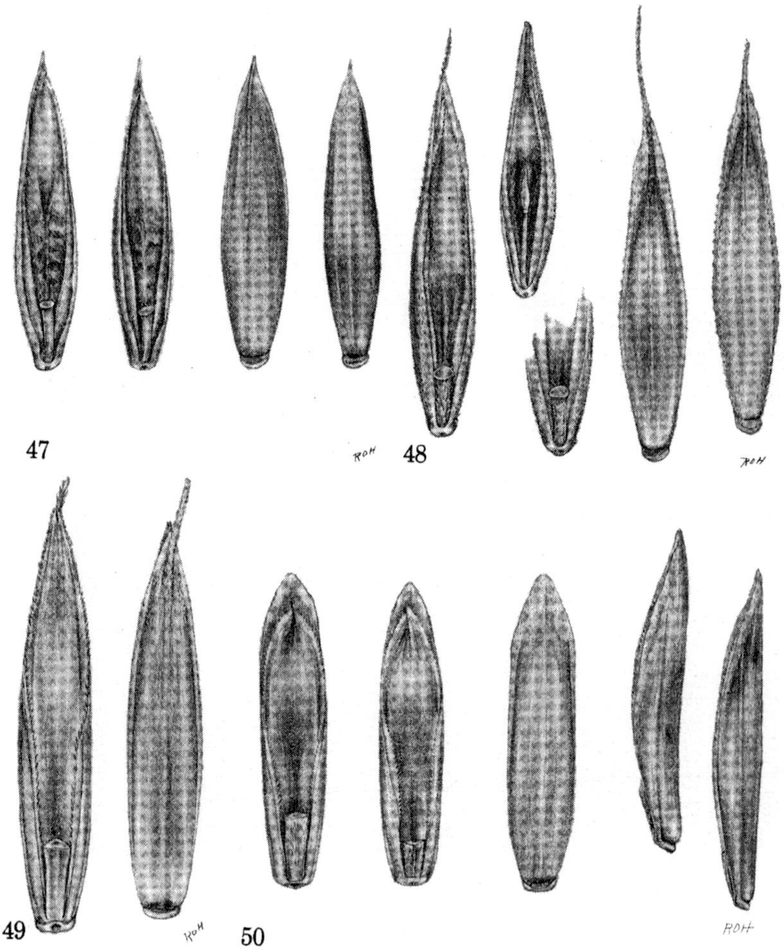

Festuca (47–48), *Lolium* (49–50), × 6

47. *F. elatior*, meadow fescue
48. *F. arundinacea*, tall fescue
49. *L. persicum*, Persian ryegrass
50. *L. perenne*, perennial ryegrass

PLATE XIX

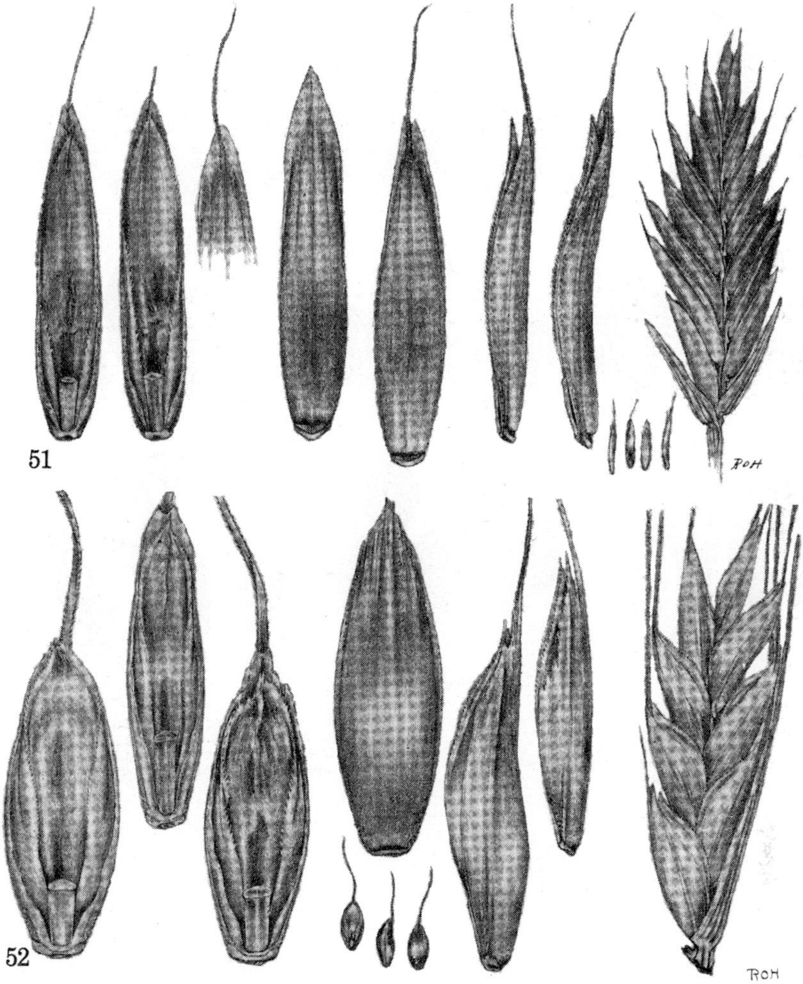

Lolium (51–52), × 6

51. *L. multiflorum*, Italian ryegrass

52. *L. temulentum*, darnel

PLATE XX

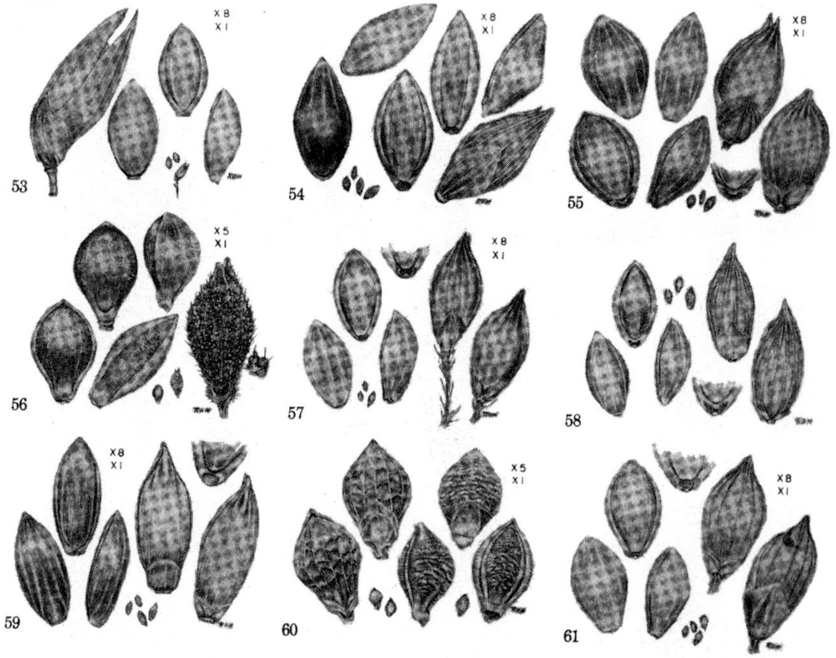

Panicum (53–61), × 25

53. *P. anceps*, beaked panicum
54. *P. antidotale*, blue panicgrass
55. *P. bergii*, **Berg's panicum**
56. *P. brachyanthum* (no common name)
57. *P. capillare*, witchgrass
58. *P. coloratum*, makarikari panicum
59. *P. dichotomiflorum*, spreading witchgrass
60. *P. fasciculatum*, browntop panicum
61. *P. gattingeri*, Gattinger witchgrass

PLATE XXI

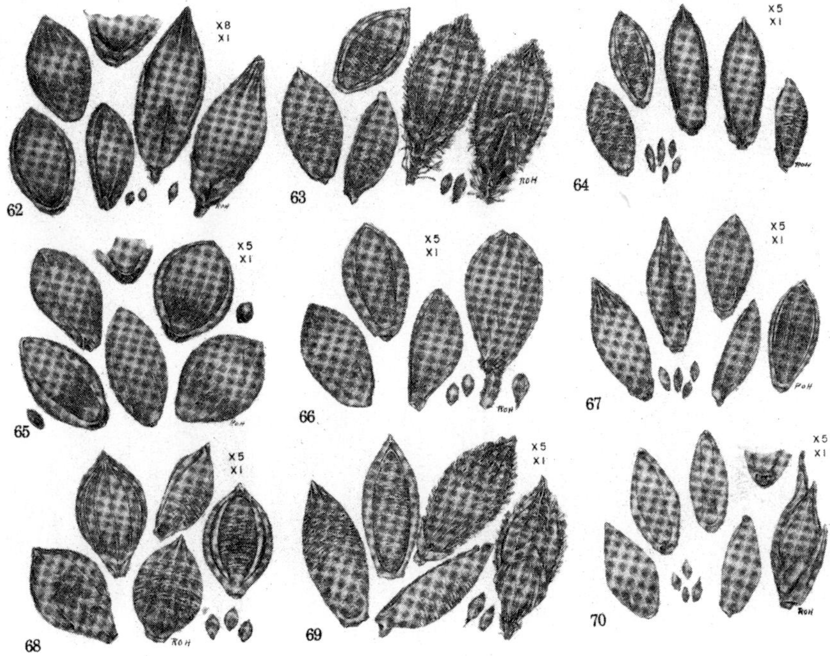

Panicum (62–70), × 25

62. *P. hillmani*, Hillman panicum
63. *P. arizonicum*, Arizona panicum
64. *P. maximum*, Guineagrass
65. *P. miliaceum*, proso millet
66. *P. obtusum*, vine-mesquite
67. *P. purpurascens*, Paragrass
68. *P. ramosum*, browntop millet
69. *P. texanum*, Texas millet
70. *P. virgatum*, switchgrass

PLATE XXII

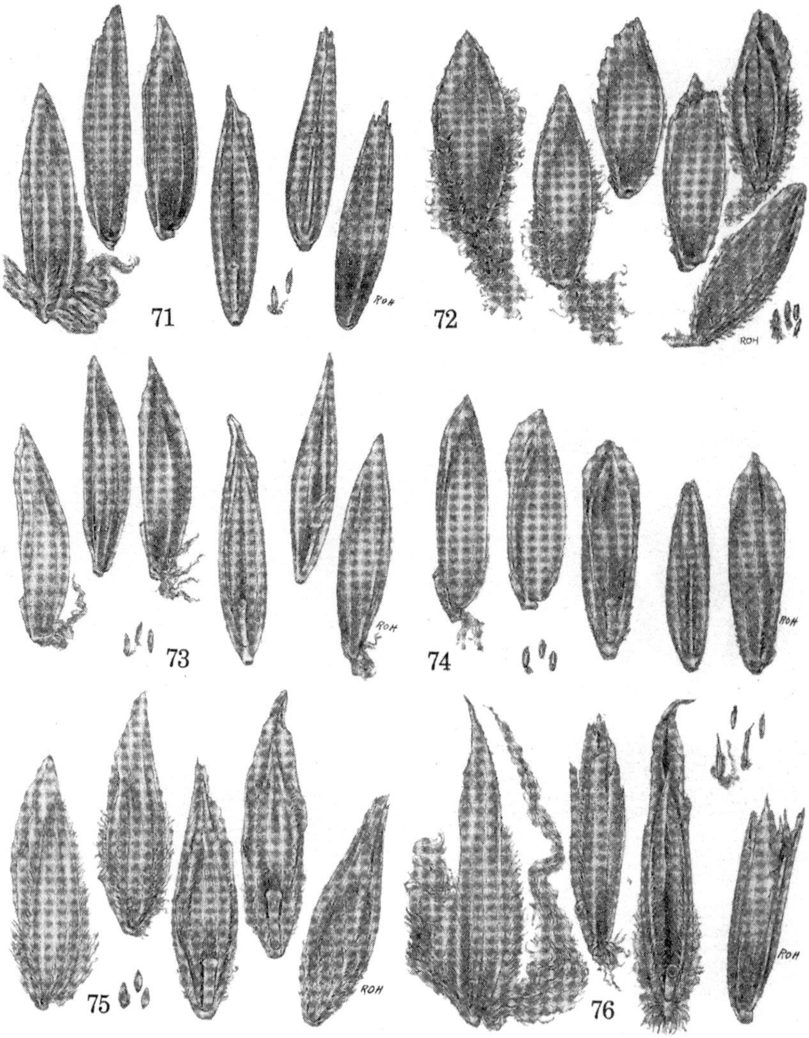

Poa (71–76), × 10

71. *P. pratensis*, Kentucky bluegrass
72. *P. pratensis*, Kentucky bluegrass var. Merion
73. *P. trivialis*, rough bluegrass
74. *P. compressa*, Canada bluegrass
75. *P. annua*, annual bluegrass
76. *P. arachnifera*, Texas bluegrass

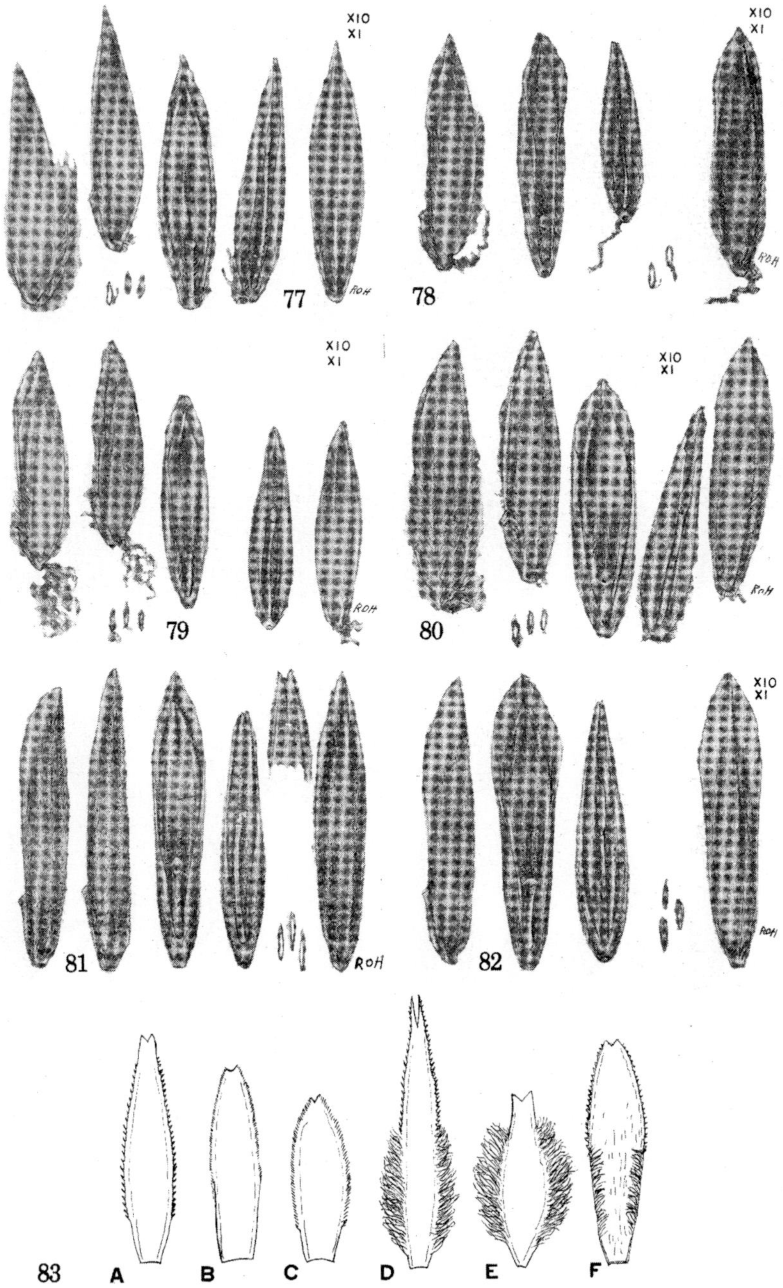

Poa (77–83), × 10

77. *P. nemoralis,* wood bluegrass
78. *P. interior,* inland bluegrass
79. *P. palustris,* fowl bluegrass
80. *P. arida,* plains bluegrass
81. *P. juncifolia,* alkali bluegrass
82. *P. ampla,* big bluegrass
83. Paleas characteristic of six species of bluegrass, greatly enlarged:
    (A) *P. pratensis* (including the var. Merion)
    (B) *P. trivialis*
    (C) *P. compressa*
    (D) *P. arachnifera*
    (E) *P. annua*
    (F) *P. arida*

PLATE XXIV

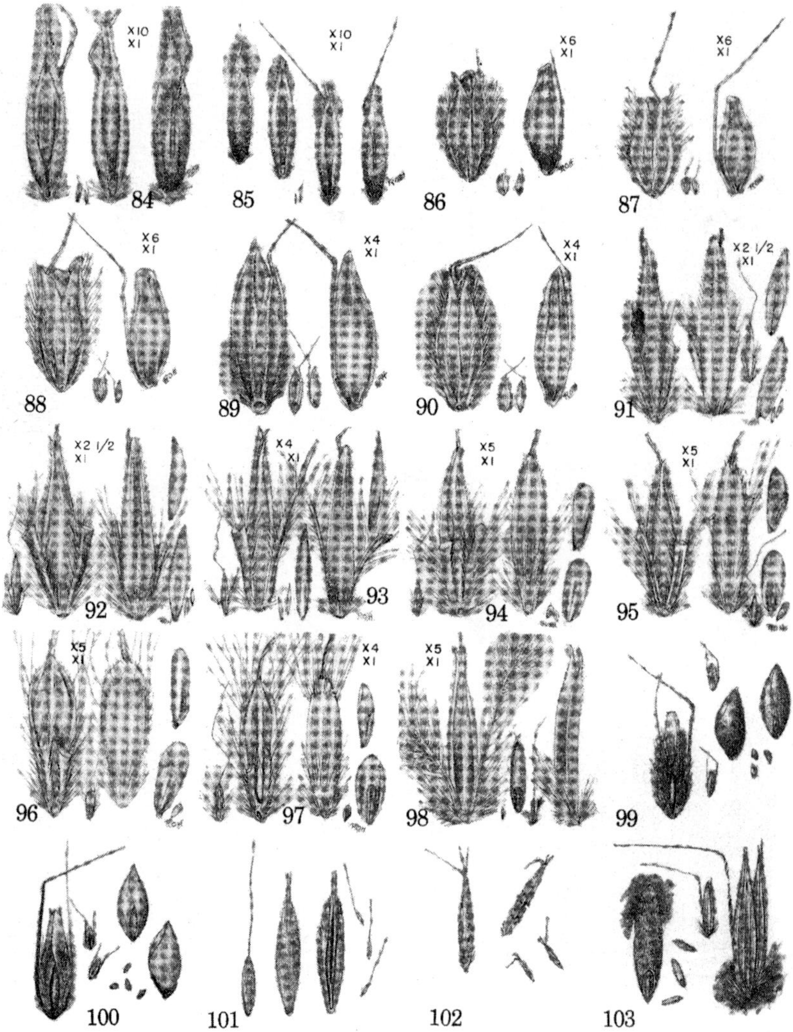

Gramineae (84–103)

84. ***Aira caryophyllea***, silver hairgrass
85. ***Aira elegans***, fine hairgrass
86. ***Alopecurus aequalis***, short-awn foxtail
87. ***Alopecurus carolinianus***, Carolina foxtail
88. ***Alopecurus geniculatus***, water foxtail
89. ***Alopecurus myosuroides***, slender foxtail
90. ***Alopecurus pratensis***, meadow foxtail
91. ***Andropogon gerardi***, big bluestem
92. ***Andropogon hallii***, sand bluestem
93. *Andropogon scoparius*, little bluestem
94. *Andropogon ischaemum*, yellow bluestem
95. *Andropogon intermedius*, Australian bluestem
96. *Andropogon annulatus*, Diaz bluestem
97. *Andropogon sericeus*, silky bluestem
98. *Andropogon virginicus*, broomsedge
99. *Anthoxanthum odoratum*, sweet vernalgrass
100. *Anthoxanthum aristatum*, annual vernalgrass
101. *Apera spica-venti*, windgrass
102. *Aristida dichotoma*, poverty three-awn
103. *Arrhenatherum elatius*, tall oatgrass

PLATE XXV

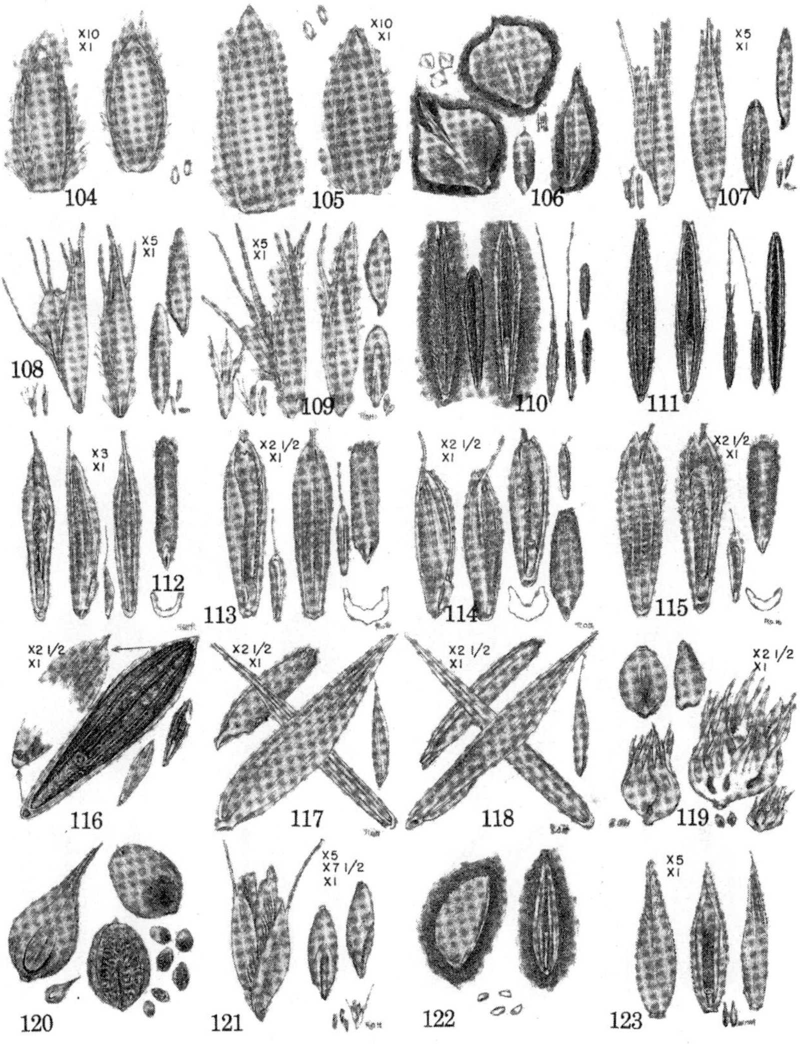

Gramineae (104–123)

104. *Axonopus affinis*, carpetgrass (narrow-leaved)
105. *Axonopus compressus*, broad-leaved carpetgrass
106. *Beckmannia syzigachne*, American sloughgrass
107. *Bouteloua curtipendula*, side-oats grama
108. *Bouteloua gracilis*, blue grama
109. *Bouteloua hirsuta*, hairy grama
110. *Bromus tectorum*, downy chess
111. *Bromus sterilis*, barren chess
112. *Bromus japonicus*, Japanese chess
113. *Bromus commutatus*, hairy chess
114. *Bromus secalinus*, chess
115. *Bromus mollis*, soft chess
116. *Bromus inermis*, smooth brome
117. *Bromus catharticus*, rescuegrass
118. *Bromus marginatus*, mountain brome
119. *Buchloë dactyloides*, buffalograss
120. *Cenchrus pauciflorus*, field sandbur
121. *Chloris gayana*, rhodesgrass
122. *Cynodon dactylon*, bermudagrass
123. *Cynosurus cristatus*, crested dogtail

PLATE XXVI

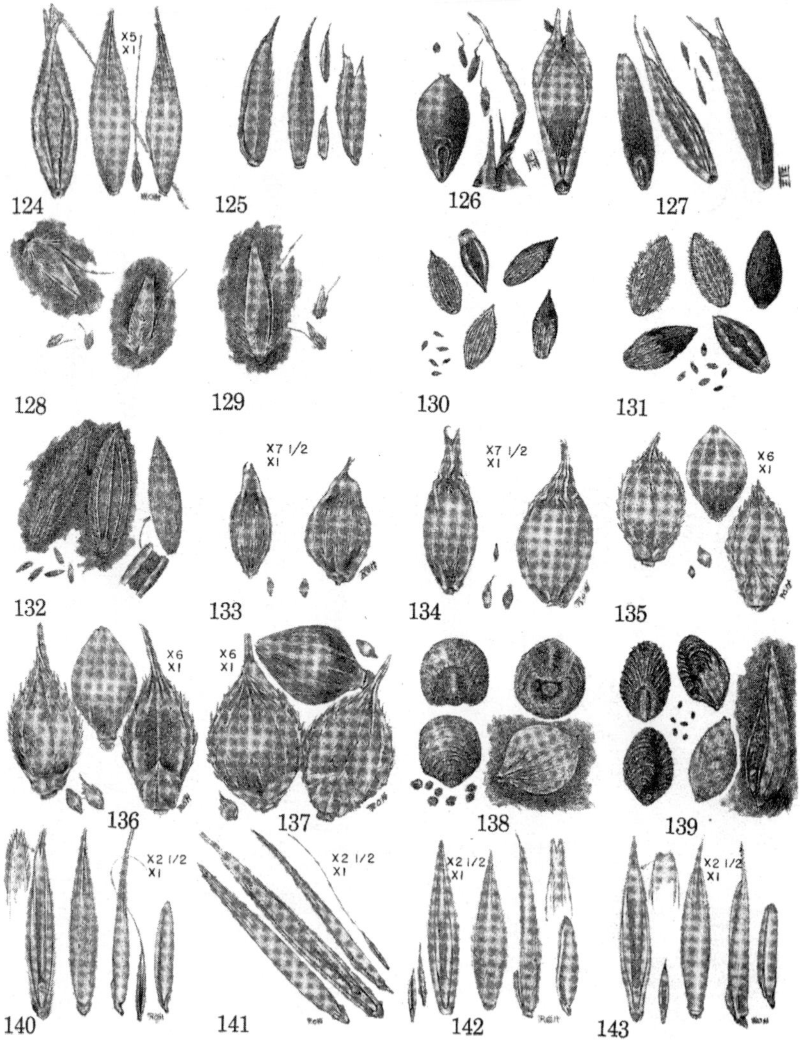

Gramineae (124–143)

124. *Cynosurus echinatus*, bristly dogtail
125. *Dactylis glomerata*, orchardgrass
126. *Danthonia spicata*, poverty oatgrass
127. *Danthonia spicata* (cleistogene)
128. *Deschampsia caespitosa*, tufted hairgrass
129. *Deschampsia flexuosa*, crinkled hairgrass
130. *Digitaria filiformis*, slender crabgrass
131. *Digitaria ischaemum*, smooth crabgrass
132. *Digitaria sanguinalis*, large crabgrass
133. *Distichlis spicata*, seashore saltgrass
134. *Distichlis stricta*, desert saltgrass
135. *Echinochloa colonum*, junglerice
136. *Echinochloa crusgalli*, barnyardgrass
137. *E. crusgalli* var. *frumentacea*, Japanese millet
138. *Eleusine coracana*, African millet
139. *Eleusine indica*, goosegrass
140. *Elymus canadensis*, Canada wild-rye
141. *Elymus glaucus*, blue wild-rye
142. *Elymus junceus*, Russian wild-rye
143. *Elymus virginicus*, Virginia wild-rye

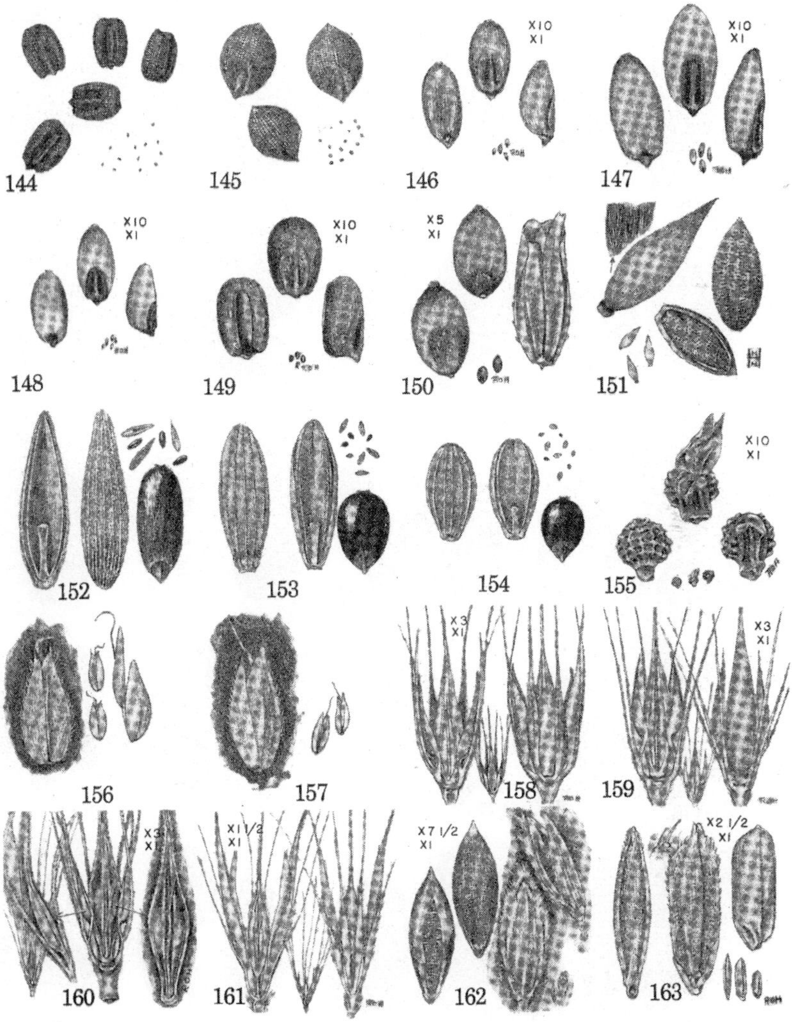

Gramineae (144–163)

144. *Eragrostis capillaris*, lacegrass
145. *Eragrostis cilianensis*, stinkgrass
146. *Eragrostis chloromelas*, Boer lovegrass
147. *Eragrostis curvula*, weeping lovegrass
148. *Eragrostis lehmanniana*, Lehmann lovegrass
149. *Eragrostis trichodes*, sand lovegrass
150. *Eremochloa ophiuroides*, centipede grass
151. *Eriochloa punctata*, Louisiana cupgrass
152. *Glyceria fluitans*, mannagrass
153. *Glyceria grandis*, American mannagrass
154. *Glyceria striata*, fowl mannagrass
155. *Hackelochloa granularis*, lizard-tail
156. *Holcus lanatus*, velvetgrass
157. *Holcus mollis*, German velvetgrass
158. *Hordeum pusillum*, little barley
159. *Hordeum brachyantherum*, meadow barley
160. *Hordeum jubatum*, foxtail barley
161. *Hordeum leporinum*, wall barley
162. *Leptoloma cognatum*, fall witchgrass
163. *Oryza sativa*, rice

PLATE XXVIII

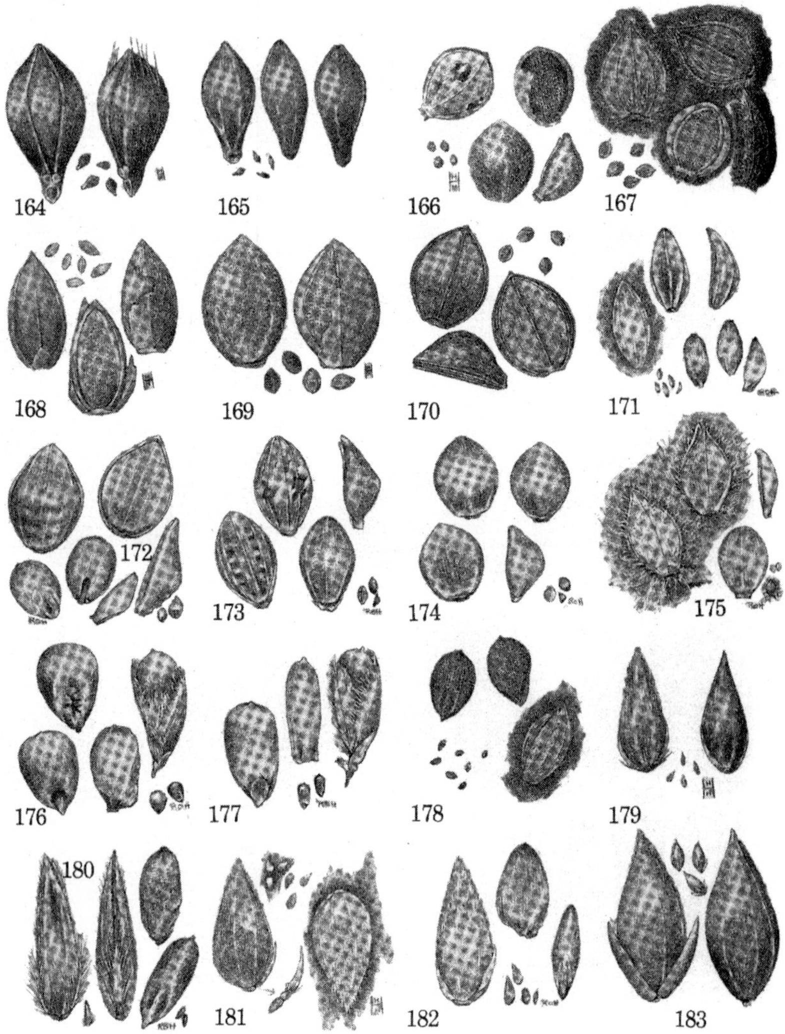

Gramineae (164–183)

164. *Oryzopsis hymenoides*, Indian ricegrass
165. *Oryzopsis miliacea*, smilo
166. *Paspalum boscianum*, bull paspalum
167. *Paspalum dilatatum*, dallisgrass
168. *Paspalum distichum*, wiregrass
169. *Paspalum floridanum*, Florida paspalum
170. *Paspalum laeve*, field paspalum
171. *Paspalum malacophyllum*, ribbed paspalum
172. *Paspalum notatum*, bahiagrass
173. *Paspalum plicatulum*, brownseed paspalum
174. *Paspalum setaceum*, slender paspalum
175. *Paspalum urvillei*, vaseygrass
176. *Pennisetum glaucum*, pearl millet
177. *Pennisetum purpureum*, napiergrass
178. *Phleum pratense*, timothy
179. *Phalaris angusta*, timothy canarygrass
180. *Phalaris arundinacea*, reed canarygrass
181. *Phalaris minor*, little canarygrass
182. *Phalaris tuberosa* var. *stenoptera*, hardinggrass
183. *Phalaris canariensis*, canarygrass

PLATE XXIX

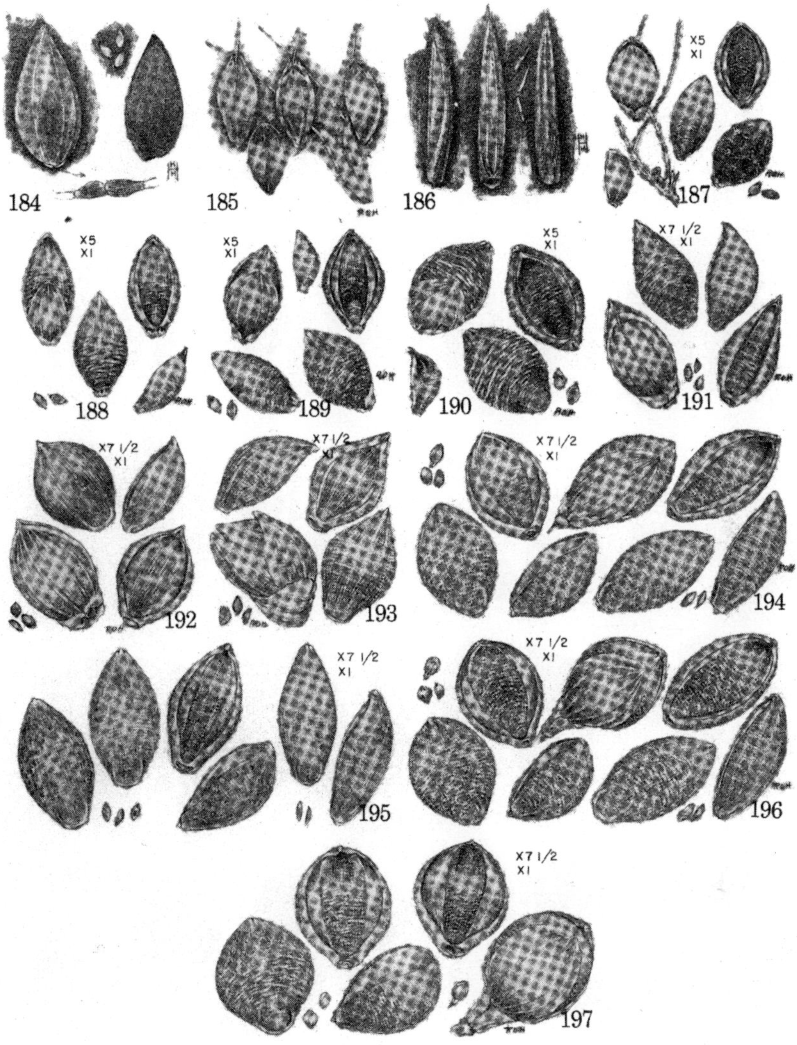

Gramineae (184–197)

184. *Phalaris paradoxa*, hood canarygrass
185. *Polypogon monspeliensis*, rabbitfoot grass
186. *Schedonnardus paniculatus*, tumblegrass
187. *Setaria viridis*, green foxtail
188. *Setaria geniculata*, knotroot bristlegrass
189. *Setaria faberi*, giant foxtail
190. *Setaria lutescens*, yellow foxtail
191. *Setaria grisebachii*, Grisebach bristlegrass
192. *Setaria magna*, giant bristlegrass
193. *Setaria macrostachya*, plains bristlegrass
194. *Setaria italica*, common foxtail millet
195. *Setaria italica*, Hungarian foxtail millet
196. *Setaria italica*, German foxtail millet
197. *Setaria italica*, White Wonder foxtail millet

PLATE XXX

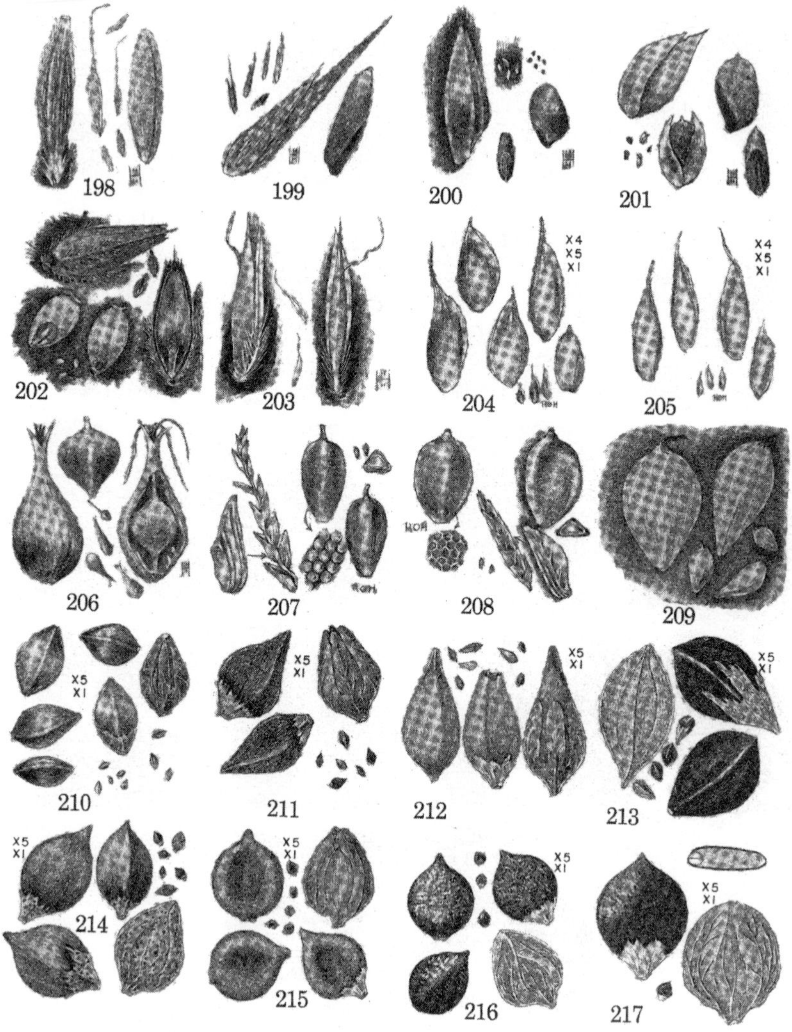

Gramineae (198–205), Cyperaceae (206–208), Liliaceae (209), Polygonaceae (210–217)

198. *Stipa viridula*, green needlegrass
199. *Sporobolus clandestinus*, scratch dropseed
200. *Sporobolus cryptandrus*, sand dropseed
201. *Sporobolus neglectus*, puff dropseed
202. *Tridens flavus*, purpletop
203. *Trisetum flavescens*, yellow oatgrass
204. *Zoysia japonica*, Japanese lawngrass
205. *Zoysia matrella*, manilagrass
206. *Carex trichocarpa*, hairyseed sedge
207. *Cyperus esculentus*, yellow nutgrass
208. *Cyperus rotundus*, nutgrass
209. *Allium vineale*, wild garlic
210. *Polygonum argyrocoleon*, silver-sheath knotweed
211. *Polygonum aviculare*, knotweed
212. *Polygonum aviculare* (cleistogenes)
213. *Polygonum convolvulus*, black bindweed
214. *Polygonum hydropiper*, common smartweed
215. *Polygonum lapathifolium*, pale ladysthumb
216. *Polygonum persicaria*, ladysthumb
217. *Polygonum pensylvanicum*, bigseed ladysthumb

PLATE XXXI

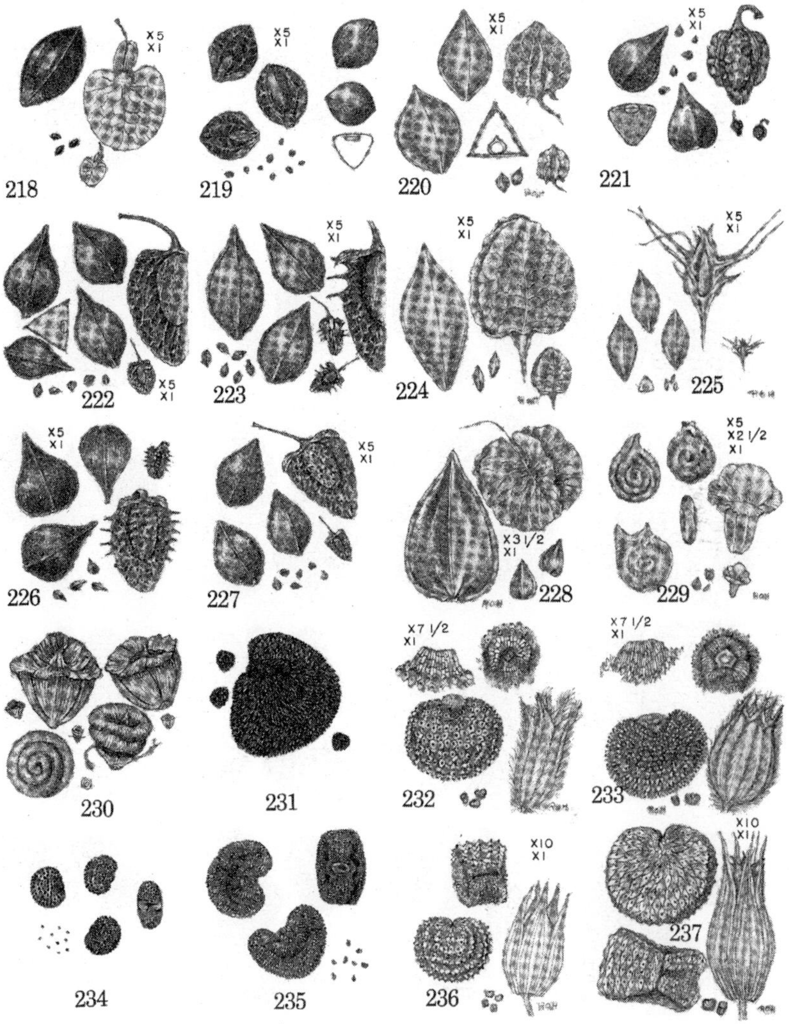

Polygonaceae (218–228), Chenopodiaceae (229–230), Caryophyllaceae (231–237)

218. *Rumex acetosa*, garden sorrel
219. *Rumex acetosella*, sheep sorrel
220. *Rumex altissimus*, peach-leaved dock
221. *Rumex conglomeratus*, clustered dock
222. *Rumex crispus*, curly dock
223. *Rumex obtusifolius*, broadleaf dock
224. *Rumex occidentalis*, western dock
225. *Rumex persicarioides*, golden dock
226. *Rumex pulcher*, fiddleleaf dock
227. *Rumex salicifolius*, willow-leaved dock
228. *Rumex venosus*, veiny dock
229. *Halogeton glomeratus*, halogeton
230. *Salsola kali* var. *tenuifolia*, Russian thistle
231. *Agrostemma githago*, corncockle
232. *Lychnis alba*, white cockle
233. *Lychnis dioica*, red campion
234. *Silene antirrhina*, sleepy catchfly
235. *Silene gallica*, English catchfly
236. *Silene conica*, conical catchfly
237. *Silene conoidea* (no common name)

PLATE XXXII

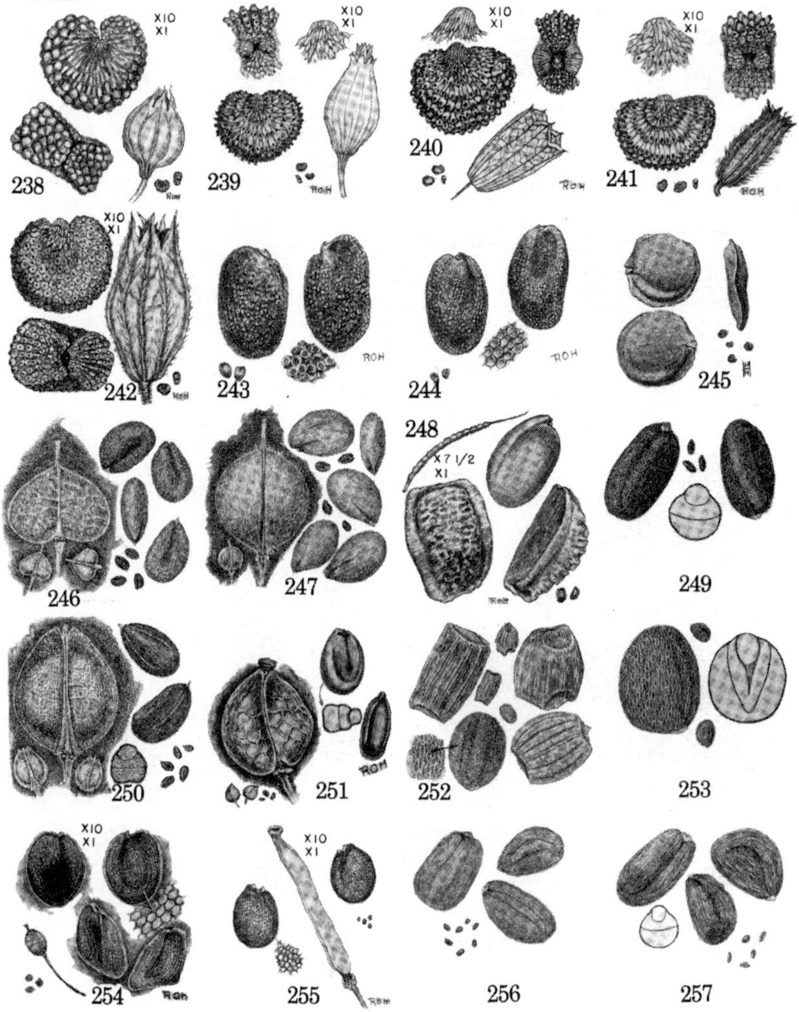

Caryophyllaceae (238–242), Cruciferae (243–257)

238. *Silene cretica*, Cretian catchfly
239. *Silene cserei*, smooth catchfly
240. *Silene cucubalus*, bladder campion
241. *Silene dichotoma*, forked catchfly
242. *Silene noctiflora*, night-flowering catchfly
243. *Barbarea verna*, early wintercress
244. *Barbarea vulgaris*, bitter wintercress
245. *Berteroa incana*, hoary alyssum
246. *Cardaria draba*, hoary cress
247. *Cardaria pubescens*, ballcress
248. *Chorispora tenella*, blue mustard
249. *Conringia orientalis*, hares-ear-mustard
250. *Lepidium campestre*, field cress
251. *Lepidium latifolium*, perennial peppergrass
252. *Raphanus raphanistrum*, wild radish
253. *Raphanus sativus*, radish
254. *Rorippa austriaca*, yellow fieldcress
255. *Rorippa sylvestris*, creeping yellowcress
256. *Sisymbrium altissimum*, tumblemustard
257. *Sisymbrium officinale*, hedgemustard

PLATE XXXIII

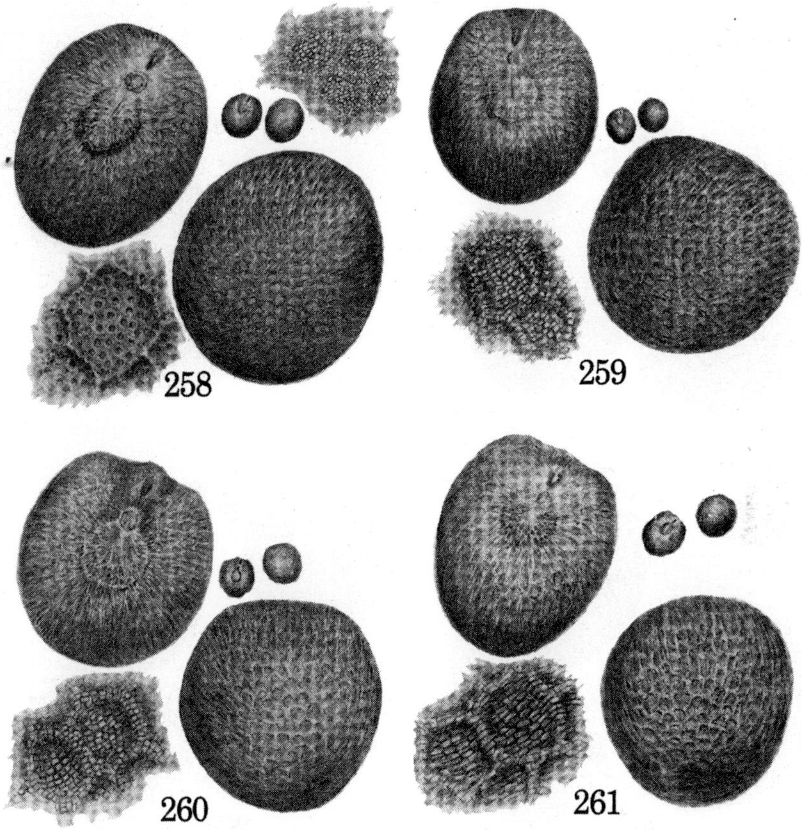

Cruciferae (258–261)

Brassica

258. *B. napus* var. *biennis*, winter rape
259. *B. napus* var. *annua*, summer rape
260. *B. napus* var. *pabularia*, Siberian kale
261. *B. napus* var. *napobrassica*, rutabaga

600705 O - 63 - 14

PLATE XXXIV

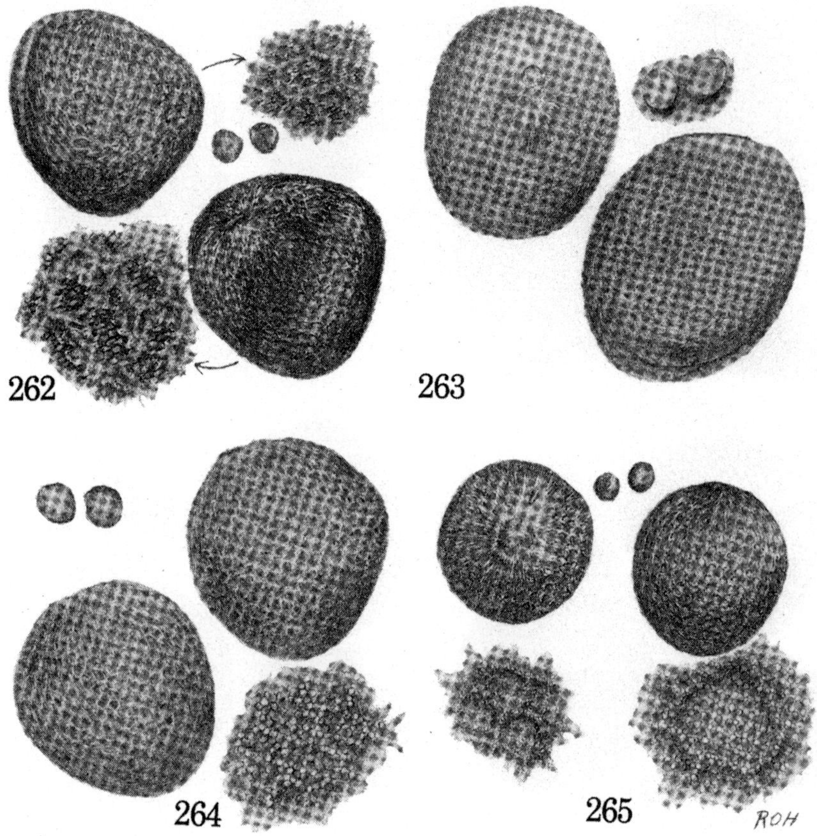

Cruciferae (262–265)

Brassica

262. *B. oleracea*, cabbage and allies
263. *B. hirta*, white mustard
264. *B. rapa* var. Shogoin turnip (or *B. campestris* var. *rapa*)
265. *B. campestris*, including turnip (*B. rapa*) and the turnip-rapes

PLATE XXXIX

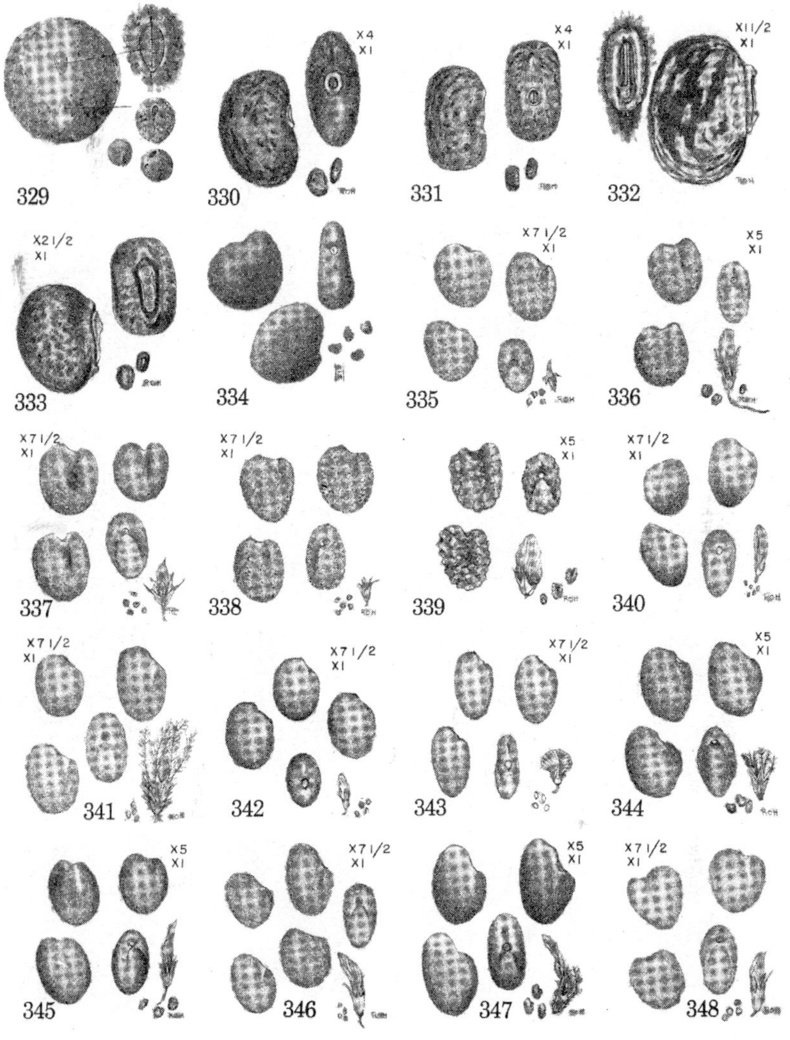

Leguminosae (329–348)

329. *Pisum sativum* var. *arvense*, field pea
330. *Pueraria thunbergiana*, kudzu
331. *Sesbania exaltata*, sesbania
332. *Stizolobium deeringianum*, velvetbean
333. *Strophostyles leiosperma*, smooth-seeded wild-bean
334. *Swainsona salsula*, Austrian peaweed
335. *Trifolium glomeratum*, cluster clover
336. *Trifolium reflexum*, buffalo clover
337. *Trifolium carolinianum*, Carolina clover
338. *Trifolium parviflorum*, teasel clover
339. *Trifolium depauperatum*, poverty clover
340. *Trifolium agrarium*, yellow hop clover
341. *Trifolium arvense*, rabbitfoot clover
342. *Trifolium dubium*, suckling clover
343. *Trifolium campestre*, large hop clover
344. *Trifolium lappaceum*, lappa clover
345. *Trifolium michelianum*, big-flowered clover
346. *Trifolium nigrescens*, ball clover
347. *Trifolium pratense*, red clover
348. *Trifolium repens*, white clover

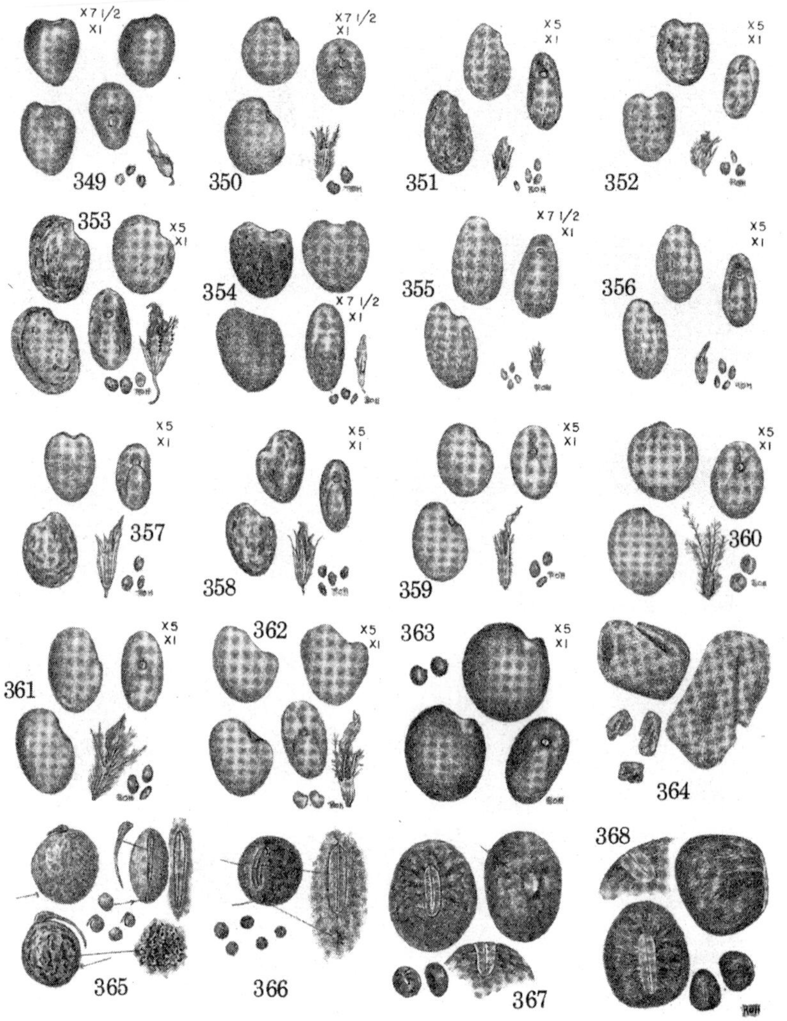

Leguminosae (349–368)

349. *Trifolium resupinatum*, Persian clover
350. *Trifolium striatum*, striate clover
351. *Trifolium bifidum*, pinole clover
352. *Trifolium fragiferum*, strawberry clover
353. *Trifolium gracilentum*, pinpoint clover
354. *Trifolium hybridum*, alsike clover
355. *Trifolium microcephalum*, small-headed clover
356. *Trifolium microdon*, thimble clover
357. *Trifolium tridentatum*, tomcat clover
358. *Trifolium variegatum*, whitetip clover
359. *Trifolium alexandrinum*, berseem clover
360. *Trifolium hirtum*, rose clover
361. *Trifolium incarnatum*, crimson clover
362. *Trifolium medium*, zigzag clover
363. *Trifolium subterraneum*, subterranean clover
364. *Trigonella foenum-graecum*, fenugreek
365. *Vicia hirsuta*, tiny vetch
366. *Vicia tetrasperma*, fourseed vetch
367. *Vicia pannonica*, Hungarian vetch
368. *Vicia melanops*, black vetch

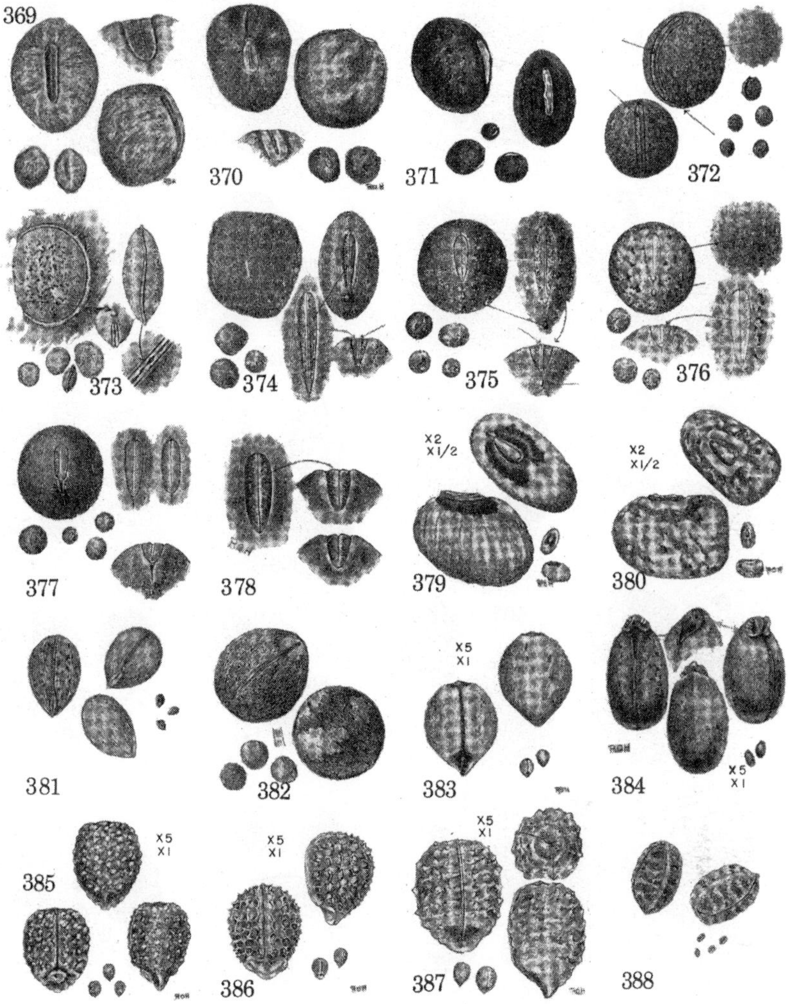

Leguminosae (369–380), Euphorbiaceae (381–388)

369. *Vicia lutea*, yellow vetch
370. *Vicia hybrida* (no common name)
371. *Vicia atropurpurea*, purple vetch
372. *Vicia cracca*, cow vetch
373. *Vicia grandiflora*, showy vetch
374. *Vicia sativa*, common vetch
375. *Vicia angustifolia*, narrowleaf vetch
376. *Vicia angustifolia*, narrowleaf vetch
377. *Vicia villosa*, hairy vetch
378. *Vicia dasycarpa*, woollypod vetch
379. *Vigna sinensis*, cowpea var. Blackeye
380. *Vigna sinensis*, cowpea var. Brabham
381. *Acalypha virginica*, three-seeded mercury
382. *Croton capitatus*, woolly croton
383. *Euphorbia corollata*, flowering spurge
384. *Euphorbia esula*, leafy spurge
385. *Euphorbia dentata*, toothed spurge
386. *Euphorbia helioscopia*, sun spurge
387. *Euphorbia marginata*, snow-on-the-mountain
388. *Euphorbia nutans*, nodding spurge

PLATE XLII

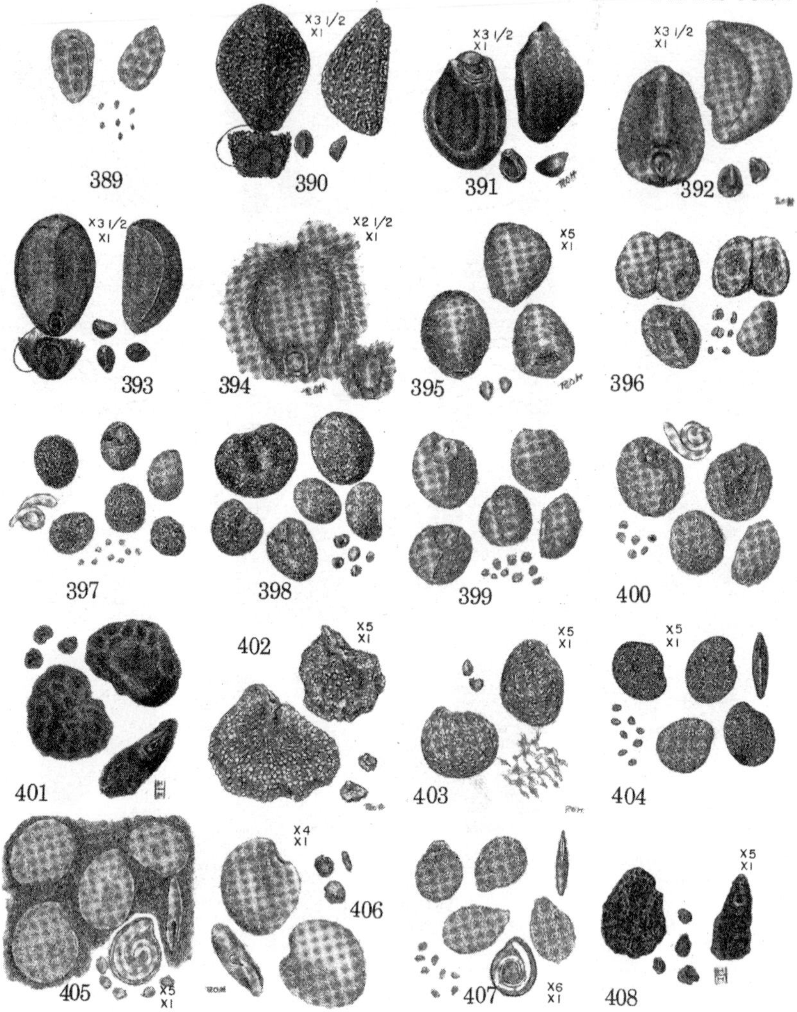

Euphorbiaceae (389), Convolvulaceae (390–395), Cuscutaceae (396–400), Solanaceae (401–408)

389. *Euphorbia supina*, spotted spurge
390. *Convolvulus arvensis*, field bindweed
391. *Convolvulus sepium*, hedge bindweed
392. *Ipomoea hederacea*, ivyleaf morning-glory
393. *Ipomoea purpurea*, common morning-glory
394. *Ipomoea pandurata*, clustered blue morning-glory
395. *Jacquemontia tamnifolia*, hairy morning-glory
396. *Cuscuta epilinum*, flax dodder
397. *Cuscuta epithymum*, clover dodder
398. *Cuscuta indecora*, large-seeded alfalfa dodder
399. *Cuscuta pentagona*, field dodder
400. *Cuscuta racemosa* var. *chiliana*, Chilean dodder
401. *Datura stramonium*, jimsonweed
402. *Physalis lobata*, purple-flowered groundcherry
403. *Physalis longifolia*, perennial groundcherry
404. *Physalis subglabrata*, taperleaf groundcherry
405. *Solanum carolinense*, horsenettle
406. *Solanum elaeagnifolium*, silverleaf nightshade
407. *Solanum nigrum*, black nightshade
408. *Solanum rostratum*, buffalo-bur

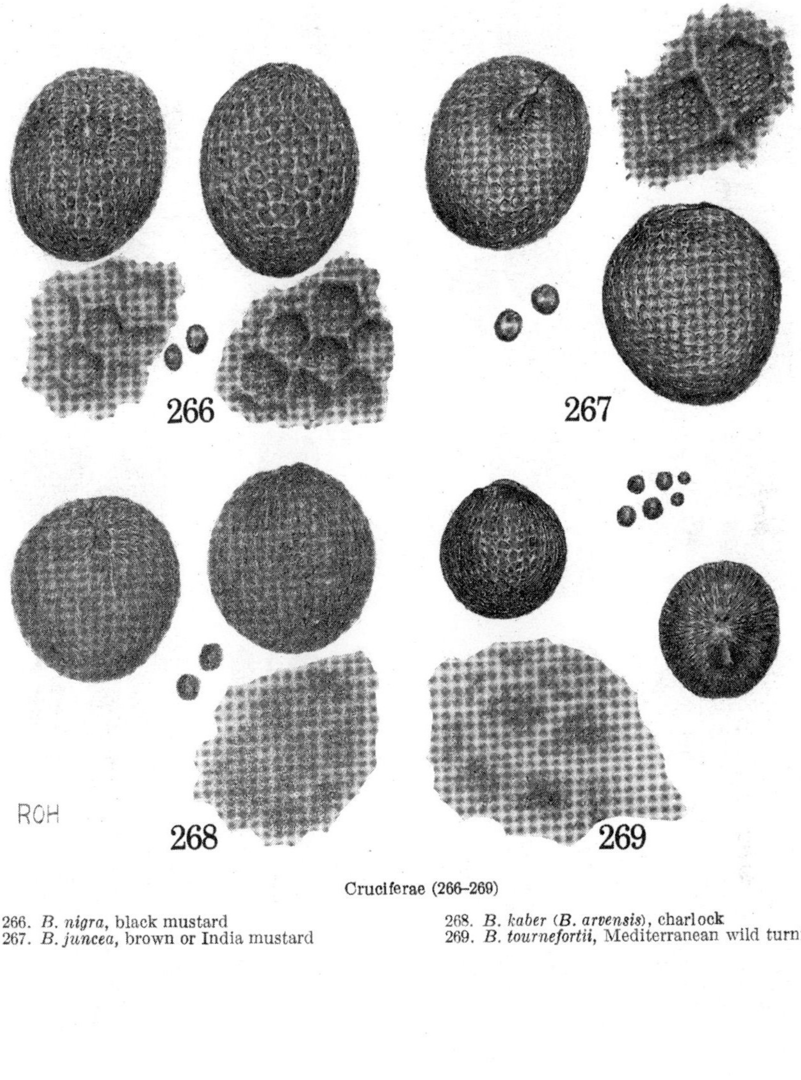

Cruciferae (266–269)

266. *B. nigra*, black mustard
267. *B. juncea*, brown or India mustard
268. *B. kaber* (*B. arvensis*), charlock
269. *B. tournefortii*, Mediterranean wild turnip

PLATE XXXVI

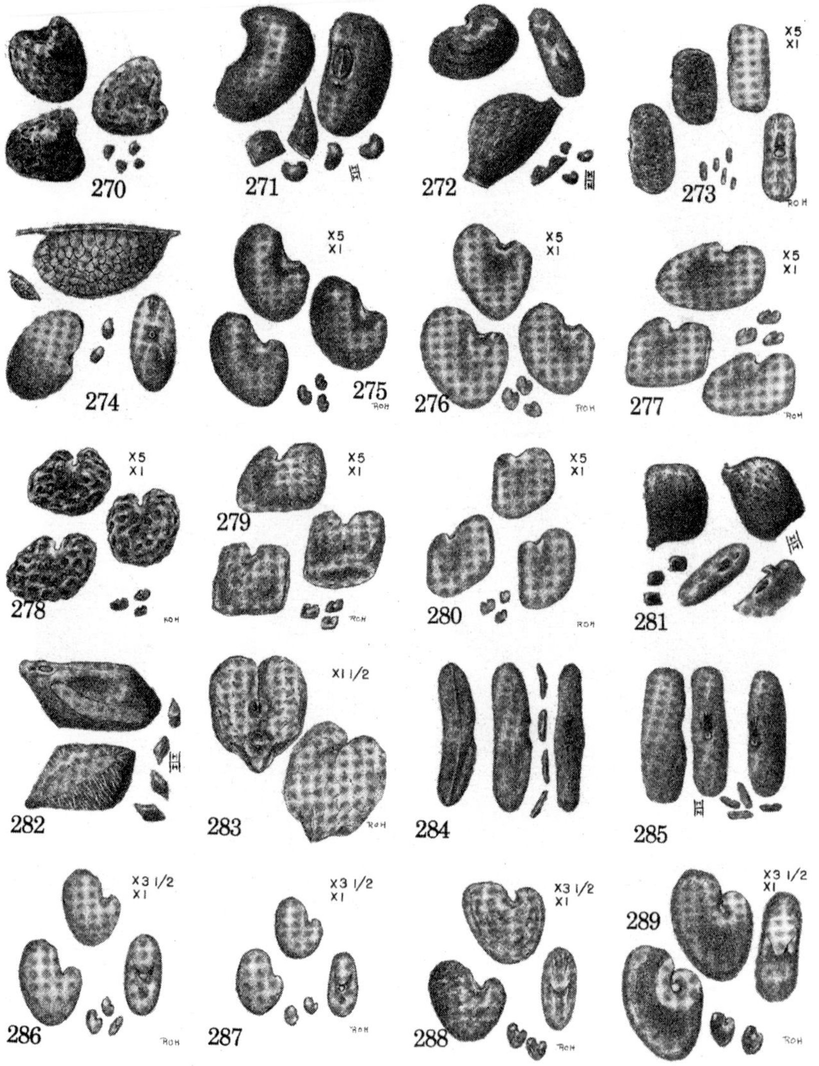

Leguminosae (270-289)

270. *Adesmia muricata*, adesmia
271. *Aeschynomene virginica*, northern jointvetch
272. *Alhagi pseudalhagi*, camelthorn
273. *Alysicarpus vaginalis*, alyceclover
274. *Anthyllis vulneraria*, kidneyvetch
275. *Astragalus chinensis*, Chinese milkvetch
276. *Astragalus cicer*, chickpea milkvetch
277. *Astragalus falcatus*, sicklepod milkvetch
278. *Astragalus flexuosus*, flexuous milkvetch
279. *Astragalus nuttallianus*, Nuttall milkvetch
280. *Astragalus rubyi*, ruby milkvetch
281. *Cassia nictitans*, sensitive pea
282. *Cassia tora*, sicklepod
283. *Cicer arietinum*, chickpea
284. *Coronilla scorpioides*, scorpion crownvetch
285. *Coronilla varia*, crownvetch
286. *Crotalaria intermedia*, slenderleaf crotalaria
287. *Crotalaria lanceolata*, lance crotalaria
288. *Crotalaria mucronata*, striped crotalaria
289. *Crotalaria speciabilis*, showy crotalaria

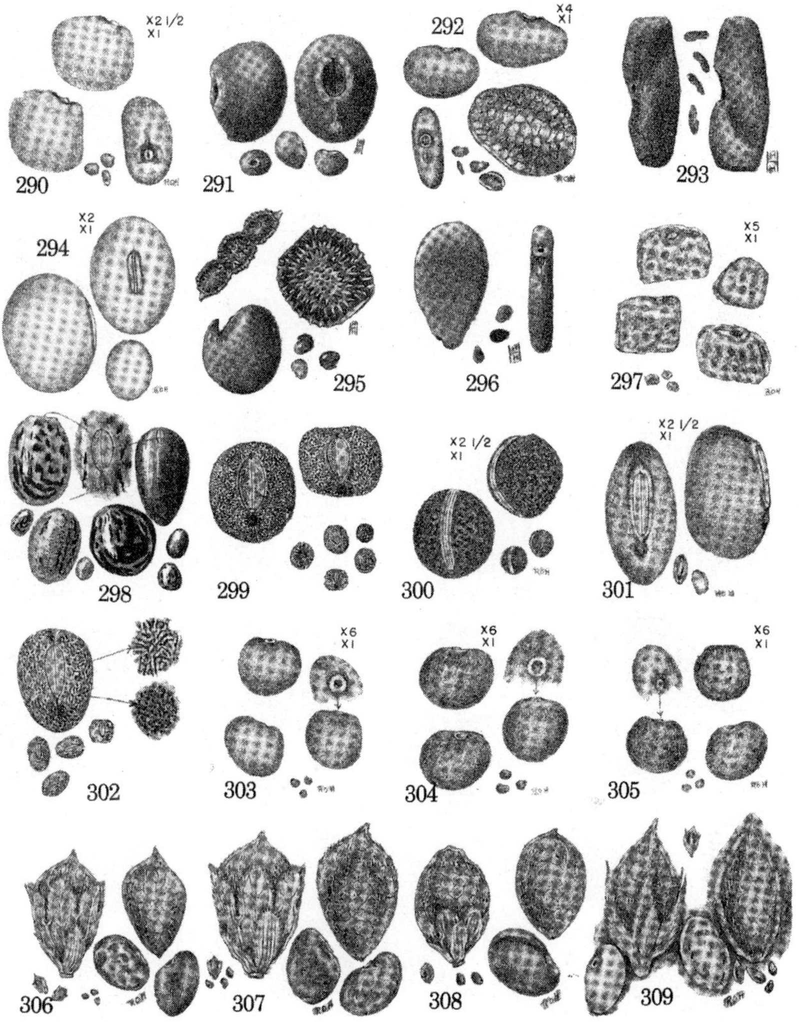

Leguminosae (290–309)

290. *Cyamopsis tetragonolobus*, guar
291. *Daubentonia texana*, rattlebox
292. *Desmodium tortuosum*, Florida beggarweed
293. *Galega officinalis*, galega
294. *Glycine max*, soybean
295. *Hedysarum coronarium*, sulla
296. *Hoffmannseggia densiflora*, Indian rushpea
297. *Indigofera hirsuta*, hairy indigo
298. *Lathyrus aphaca*, yellow vetchling
299. *Lathyrus hirsutus*, roughpea
300. *Lathyrus sylvestris*, flatpea
301. *Lathyrus tingitanus*, Tangier pea
302. *Lathyrus tuberosus*, groundnut pea
303. *Lespedeza striata*, Common striate lespedeza
304. *Lespedeza striata*, Kobe striate lespedeza
305. *Lespedeza stipulacea*, Korean lespedeza
306. *Lespedeza cuneata*, sericea lespedeza
307. *Lotus angustissimus*, slenderpod deervetch
308. *Lotus corniculatus*, birdsfoot trefoil
309. *Lotus hispidus*, hispid deervetch

PLATE XXXVIII

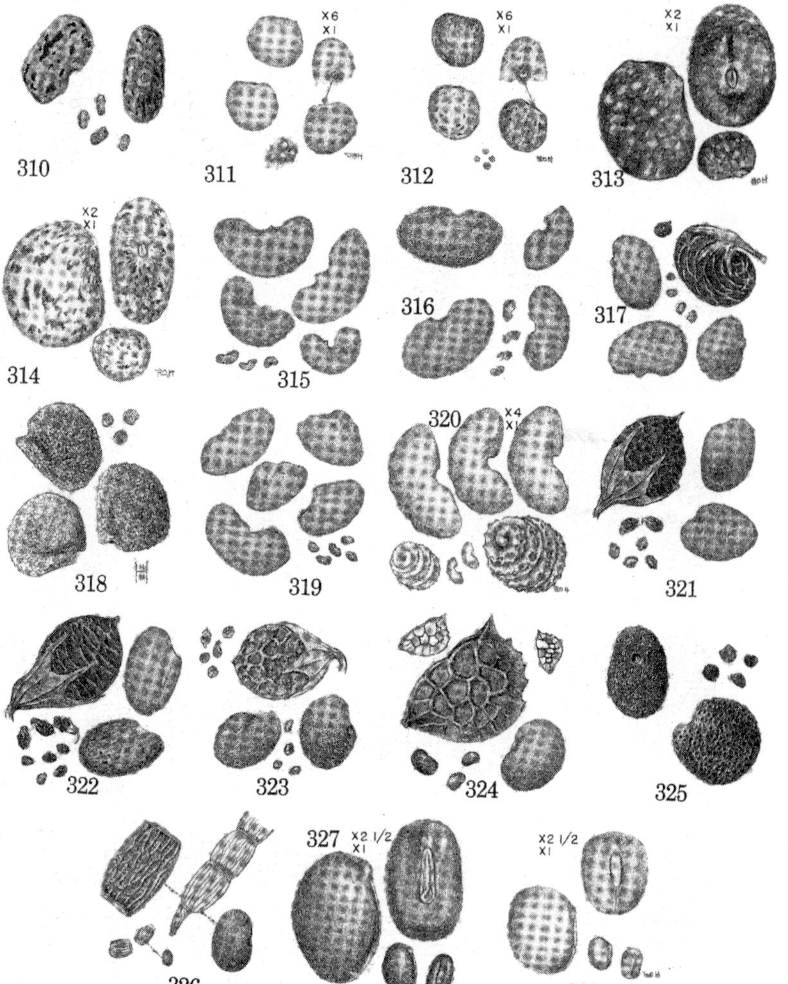

Leguminosae (310–328)

310. *Lotus purshianus*, prairie trefoil
311. *Lotus uliginosus* var. *glabriusculus*, smooth big trefoil
312. *Lotus uliginosus* var. *villosus*, hairy big trefoil
313. *Lupinus angustifolius*, blue lupine
314. *Lupinus luteus*, yellow lupine
315. *Medicago arabica*, spotted bur-clover
316. *Medicago hispida*, California bur-clover
317. *Medicago lupulina*, black medic
318. *Medicago orbicularis*, button-clover
319. *Medicago sativa*, alfalfa
320. *Medicago tuberculata*, cogwheel clover
321. *Melilotus alba*, white sweetclover
322. *Melilotus officinalis*, yellow sweetclover
323. *Melilotus indica*, sourclover
324. *Onobrychis viciaefolia*, sainfoin
325. *Ononis repens*, ononis
326. *Ornithopus sativus*, serradella
327. *Phaseolus angularis*, adzuki bean
328. *Phaseolus aureus*, mung bean

PLATE XLIII

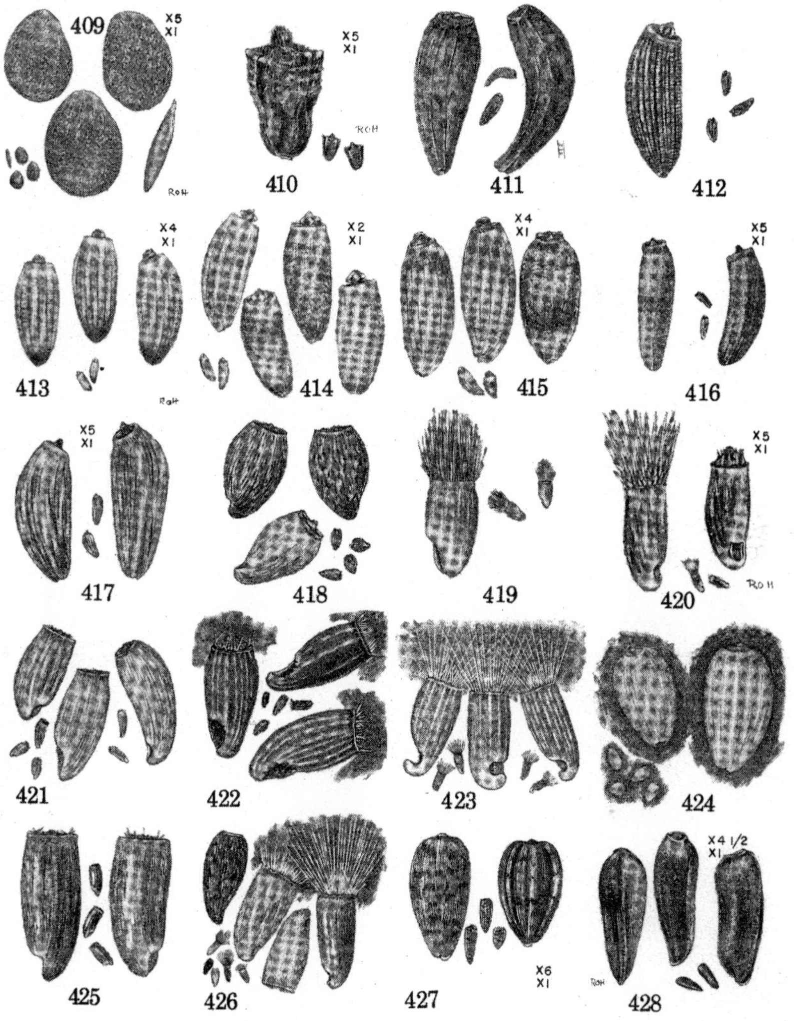

Solanaceae (409), Compositae (410–428)

409. *Solanum triflorum,* cutleaf nightshade
410. *Ambrosia psilostachya,* perennial ragweed
411. *Arctium* sp., burdock
412. *Carduus crispus,* curly thistle
413. *Carduus acanthoides,* plumeless thistle
414. *Carduus macrocephalus,* large-headed plumeless thistle
415. *Carduus pycnocephalus,* Italian thistle
416. *Cirsium arvense,* Canada thistle
417. *Cirsium vulgare,* bull thistle
418. *Centaurea calcitrapa,* purple star-thistle
419. *Centaurea cyanus,* cornflower
420. *Centaurea iberica,* Iberian star-thistle
421. *Centaurea jacea,* brownscale knapweed
422. *Centaurea maculosa,* spotted knapweed
423. *Centaurea melitensis,* Malta star-thistle
424. *Centaurea picris* (*C. repens*), Russian knapweed
425. *Centaurea scabiosa,* scabious star-thistle
426. *Centaurea solstitialis,* yellow star-thistle
427. *Helianthus annuus,* common sunflower
428. *Helianthus ciliaris,* Texas blueweed